EINSTEIN, HISTORY, AND OTHER PASSIONS

Masters of Modern Physics

Published Volumes

EINSTEIN, HISTORY, AND OTHER PASSIONS

GERALD HOLTON

American Institute of Physics

AIP Press
American Institute of Physics
500 Sunnyside Boulevard
Woodbury, NY 11797-2999

Library of Congress Cataloging-in-Publication Data

Holton, Gerald James.
    Einstein, history, and other passions / Gerald Holton.
        p.   cm.—(Masters of modern physics; v. 16)
        Includes index.
        ISBN 1-56396-333-7
        1.   Science—Miscellanea.   2.   Einstein, Albert, 1879–1955—Knowledge—
        Science.   3.   Science—Social aspects—History.   4.   Physicists—Biography.
        5. Science—Study and teaching.
        I. Title. II. Series.
        Q173.H7342   1995   95–37483
        500—dc20   CIP

This book is volume sixteen of the Masters of Modern Physics series.

10   9   8   7   6   5   4   3   2   1

Contents

Part Three: PERSONALITIES IN PHYSICS

Part Four: SCIENCE FOR STUDENTS

About the Series

Masters of Modern Physics introduces the work and thought of some of the most celebrated physicists of our day. These collected essays offer a panoramic tour of the way science works, how it affects our lives, and what it means to those who practice it. Authors report from the horizons of modern research, provide engaging sketches of friends and colleagues, and reflect on the social, economic, and political consequences of the scientific and technical enterprise.

Authors have been selected for their contributions to science and for their keen ability to communicate to the general reader—often with wit, frequently in fine literary style. All have been honored by their peers and most have been prominent in shaping debates in science, technology, and public policy. Some have achieved distinction in social and cultural spheres outside the laboratory.

Many essays are drawn from popular and scientific magazines, newspapers, and journals. Still others—written for the series or drawn from notes for other occasions—appear for the first time. Authors have provided introductions and, where appropriate, annotations. Once selected for inclusion, the essays are carefully edited and updated so that each volume emerges as a finely shaped work.

Masters of Modern Physics is edited by Robert N. Ubell and overseen by an advisory panel of distinguished physicists. Sponsored by the American Institute of Physics, a consortium of major physics societies, the series serves as an authoritative survey of the people and ideas that have shaped twentieth-century science and society.

Preface

Today's scientists pride themselves on feats of intellectual advancement they have witnessed in their lifetimes. Meanwhile modern historians and philosophers struggle to understand the dynamics by which the ever-growing flood of new knowledge seeping through the laboratory walls has fundamentally altered modern life. The old, comfortable idea of a fairly strict separation between the culture of the laboratory and that of the outside world was never quite true to the organic nature of the modern state, and in recent times its believers experienced a rude awakening.

I might be permitted an anecdote that illustrates the point. Having made contributions in applied science during World War II that became essential for defeating the nemesis of fascism, many physicists longed to return to the safety of their ivory towers, to do intellectually exciting work that might also have the merit, in G.H. Hardy's formulation, of being useless to anyone else. In my case, I had already been predisposed in that direction, having felt compelled not to follow the usual call to Los Alamos, and having been put to work instead in a laboratory developing defensive equipment (against the enemy's expected use of gas warfare). At any rate, when I resumed my graduate studies after the peace declaration, I carefully chose for my Ph.D. thesis a topic that would be at the intellectual frontier but also would have no foreseeable relationship to anything involving weapons: the molecular relaxation of liquids under high pressures, where one could hope to test rival theories of the liquid state. I vividly recall my consternation when the very first request for a copy of my results came from Los Alamos. It turned out that I had unwittingly produced data needed there for the design of liquid-cooled, pressurized nuclear reactors for submarines.

The interactions—some plainly visible, others subterranean but equally vital—between the world of science and the rest of twentieth-century culture have forced a complete reformulation of previous answers to that long list of questions that are basic to any understanding of modernity itself. How do the findings and methods of science affect the polity as a whole? How does the public react to the hopes and fears aroused by those new findings? How, indeed, does the individual scientific imagination work, and how does the communal understanding among scientists arise from the diverse individual contributions? How can science education be effective in the face of the accelerations of history and the periodic growth of anti-science attitudes? And even, how do we establish the cultural and political mandate for the pursuit of rationality and science? As Isaiah Berlin has pointed out, not one of the nineteenth-century prophets predicted that as the twentieth century progressed its culture would be more and more dominated by "the rejection of reason and order, as being prison houses of the spirit." This rejection is a rebellion against the growing power and incursions of science and technology in daily life.

In my writings, beyond contributions to physics literature, I have attempted to provide a framework for an understanding of the cultural dimension of modern science, and, where possible, to put such understanding to work. The main findings have been presented in four books: *Thematic Origins of Scientific Thought: Kepler to Einstein* (Harvard University Press, 2nd, revised edition, 1988), *The Scientific Imagination: Case Studies* (Cambridge University Press, 1978), *The Advancement of Science, and Its Burdens* (Cambridge University Press, 1986), and *Science and Anti-Science* (Harvard University Press, 1993). This new volume brings together more results from that continuing endeavor, some of them just completed, others from earlier years but reconsidered and, where necessary, rewritten.

This book is divided into four main parts. The first is devoted to the light cast by Albert Einstein's thoughts on our time and our self-understanding—part of a long-standing preoccupation. In "Einstein's Influence on the Culture of Our Time," we can document the enormous variety of his real or imagined impacts apart from science itself, in a cultural spectrum that reaches from modern painting and literature to child psychology. The next essay, "Einstein and the Goal of Science," centers on his science and epistemology, by means of the analysis of an essay from Einstein's pen which his biographer and colleague, Philipp Frank, called the "finest formulation of his views on the nature of physical theory."

Next we focus on the letters of Einstein and Mileva Marić during the early years of their deep attachment to one another. "Of Physics, Love, and Other Passions" reveals a picture of young Albert much at variance with the usual icon of the abstracted, world-weary patriarch; at the same time, it helps us understand the courageous and ultimately tragic figure of Mileva. The following chapter tries to enter the mind of the scientist when he raises the question, as Einstein put it, "What, Precisely, is 'Thinking'?"—and to answer it by using Einstein's autobiographical comments on this topic.

Part Two concerns the dynamics of science as it manifests itself in our modern civilization generally, as well as in the growth of scientific thought itself. "What Place for Science in Our Culture at the 'End of the Modern Era'?" traces the widely fluctuating levels of regard for the values assigned to science since the beginning of the twentieth century, and so might help us preview the world we are now entering. I hope it makes clear why the fragile "social contract" between science and society that has been emerging lately is so fundamentally different from that on which most scientists have been brought up.

While that essay is one of the most recent, the next, "The Public Image of Science," is based on one of my earliest; it presents a classification of the various psychological perspectives on science that may help explain the often contradictory images of science at large among intellectuals, students, and the public.

The evolution of the trustworthiness of scientific research results ("Doing One's Damnedest...") is the topic of the next essay. Although one tends to forget it, the level of trust that has been achieved so far is the product of a long and often painful evolution of ideas and inventions, from the introduction of algebra to the decision of what should be considered a "fact," from the invention of how to publish a finding to the founding of the modern laboratory, from the early statistical treatment of data to the ethos of teamwork in contemporary "Big Science."

"Imagination in Science" further counters those (largely false) public images of science presented in the sixth essay, by referring to the way a historian of science can now glimpse aspects of the scientific imagination in action—specifically the visual, the metaphorical, and the thematic imagination. What emerges is, of course, quite opposite both to the purely logical–mechanical, nineteenth-century model of the scientific mind, and equally to the allegedly impersonal–rational model which all too often serves today as a condemned straw man.

The next essay, "Understanding the History of Science," lays out in some detail the tool kit for analyzing, with professional responsibility, an "event" in the history of science. This goes well beyond the old controversy whether to prefer "internalist" or "externalist" approaches. Being skeptical of the merits of the various ambitious, overarching theories of historical change in scientific knowledge, I prefer the resolution of major events through a multi-component approach; one of the advantages is that it allows the historical and other connections between events to be more easily woven into what may be considered an "understanding" of these events.

Part Three is devoted to biographical sketches. Elsewhere I have dealt with the work of many others, including Johannes Kepler, Niels Bohr, and R. A. Millikan. Here, I am presenting my perspective on five quite different, major scientists—J. Robert Oppenheimer, Percy W. Bridgman, I.I. Rabi, Richard Feynman, and Michael Polanyi—all of whom I came to know. Their stories clearly illustrate the increasingly complex coupling between abstract research science and concrete societal demands during this century. In this sense, the events presented in Part Three test some of the ideas developed in Part Two.

Finally, Part Four deals with education in science—a topic which has remained perplexing despite intense scrutiny, but which for me, at any rate, remains a passion, together with the study of the science-society link. It was my good fortune to become involved in college science teaching in the heady days when president James B. Conant at Harvard not only helped launch the General Education approach in the university but also exhibited his earnest leadership by teaching such a course himself, and personally recruiting the interests of young scientists on the faculty. The high-level encouragement to develop additional courses that taught good science while also trying to bridge the cultural gap between the sciences and the humanities resulted in my taking the risk, as a very young physics instructor, to devote some time to the writing of a new kind of text, one that approached physics through its history and through its connections with the other sciences and cultural fields beyond. The result, clumsily entitled *Introduction to Concepts and Theories in Physical Science*, was published in 1952—and its revised edition (with Stephen G. Brush) is still in print at Princeton University Press. The book later became the basis for a national curriculum project for secondary schools, under the title *The Project Physics Course*, which is also still in use in the United States and in several foreign-language translations.

The essay, "How Can Science Courses Use the History of Science?," spells out the educational philosophy behind these related pedagogic efforts, which in their general direction seemed to me, at least as applicable to the needs of today's students as of those in the past.

"Faraday's 'Advice to a Lecturer,' Updated" deals with another, but not unrelated, pedagogic challenge: the processes by which instruction is actually conveyed in a science class, including the subliminal interactions that take place even during flamboyant demonstrations. The chapter was motivated by the lack of a satisfactory theory of how these processes work.

A chief intention of the volume as a whole is that it may, along its several dimensions, illustrate the civilizing power of thought about science. Beyond that, my hope is that these pages may indicate new answers to old questions, and new questions for old certainties.

Part One:
LEARNING FROM EINSTEIN

Einstein's Influence on the Culture of Our Time

The effects a scientific advance may have on the whole era are well illustrated in the case of Einstein's early work. One's first thoughts go of course to his influence on science itself; but then I shall also deal with the other, sometimes strange effects he is said to have had on the rest of the culture of our time.

Within physics, there was no immediate recognition of the transforming nature of his early papers. Six years elapsed after the first publication of the special theory of relativity before it had established itself sufficiently to merit a textbook (Max von Laue's *Das Relativitätsprinzip*, 1911), and for some years after that the theory continued to be confused by most scientists with the electrodynamics of H.A. Lorentz. Einstein's ideas on quantum physics, published from 1905 on, were also generally neglected or discounted for years. R. A. Millikan, on accepting his Nobel Prize for 1923, confessed that the validity of Einstein's "bold, not to say reckless" explanation of the photoelectric effect forced itself on him slowly, "after ten years of testing...[and] contrary to my own expectation." The transcripts of the questions raised in scientific meetings in the decade after 1905 demonstrated the larger intellectual effort required at the time to enter fully into the meaning of the new physics.

Today, virtually every student who wishes to can learn at least the elements of relativity or quantum physics before leaving high school, and the imprint of Einstein's work on the different areas of physical science is so large and varied that a scientist who tries to trace it would be hard put to know where to start. A modern dictionary of scientific terms contains thirty-five entries bearing his name, from "Einstein: A unit of light energy used in photochemistry" and "Einstein-Bose statistics" to "Einstein tensor" and "Einstein viscosity equation."[1] It is ironic that now, a quarter of a century after his death, there is in many branches of the physical sciences more awareness of his generative role than would have been credited during the last decade or two of his life. His ideas became essential for laying out conceptual paths for contemporary work in astronomy or cosmology, for unifying gravitation with the quantum field theory of gauge fields, or even for understanding new observations that were not possible in his time but were predicted by him (as in his 1936 paper on the optical lens formed by gravitational fields).

Apart from changing science itself, Einstein has reached into the daily life of virtually every person on the globe in direct or indirect ways through the incorporation of his ideas on physics into a vast range of technical devices and processes. I need cite only some of the most obvious ones. Every photoelectric cell can be considered one of his intellectual grandchildren. Hence, we are in his debt whenever photo emission or absorption is used, in the home or on the job, to capture an image by means of a television camera, or to project the optical soundtrack of a motion picture, or to set the page of a book or newspaper by photocomposition, or to make a telephone call over fiber cable. In each case, if a law required a label on the appliance giving its intellectual content or pedigree, such a display would list prominently: "Einstein, *Annalen der Physik* 17 (1905), pp. 132–148; 20 (1906), pp. 199–206," etc.

One would find an entry of this sort also on the laser, whose beam was used to lay out the highway on which one travels to the office or to site the building itself ("Einstein, *Physikalische Zeitschrift* 18 (1917), pp. 121–128," etc.). Or again, the same kind of answer comes if one lists key ideas that helped to make possible modern electric machinery, such as power generators, or precision clocks that allow the course of planes and ships to be charted. Einstein appears also, if one looks for the ancestry of the ideas, in quantum and statistical physics, by which solid-state devices operate, from calculators and computers to the transistor radio and the ignition system; and once more, even when one takes one's vitamin pill or other pharmaceutical drug, for it is likely that its commercial production involved diffusion processes, first explained in Einstein's papers on Brownian movement and statistical mechanics ("Einstein, *Annalen der Physik* 17 (1905), pp. 549–560," etc.).

As Edward M. Purcell has remarked, since the magnetism set up by electric currents is a strictly relativistic effect, derivable from Coulomb's law of electrostatics and the kinematics of relativity, and nothing more, it requires no elaboration to discuss "special relativity in engineering": "This is the way the world *is*. And it does not really take gigavolts or nanoseconds to demonstrate it; stepping on the starter will do it!"[2] It is not much to say that even in our most common experiences, that unworldly theoretician's publications help to explain what happens to us all day—indeed, from the moment we open our eyes on the light of the morning, since the act of seeing is initiated by a photochemical reaction ("Einstein, *Annalen der Physik* 37 (1912), pp. 832–838; 38 (1912), pp. 881–884," etc.).

The proverbial man in the street is quite blissfully ignorant of all that, and has preferred to remain so, even while expecting fully that, mysteriously yet automatically, a stream of practical, benign "spin-offs" continues from the pursuit of pure science. But the philosopher, the writer, the artist, and many others outside the scientific laboratories could not help but be caught up to some extent by the wave that spread beyond science and technology, at first slowly, then with astonishing intensity. As the best scientists were coming to understand what Einstein had done, the trumpets began to sound. Even Max Planck, a person conservative in thought and expression, enthused by 1912: "This new way of thinking...well surpasses in daring everything that has been achieved in speculative scientific research, even in the theory of knowledge...This revolution in the physical *Weltanschauung*, brought about by the relativity principle, is to be compared in scope and depth only with that caused by the introduction of the Copernican system of the world."[3] At the same time, on the other end of the philosophical spectrum, the followers and heirs of Ernst Mach rushed to embrace relativity as a model triumph of positivistic philosophy. In the inaugural session of the Gesellschaft für positivistische Philosophie in Berlin (11 November 1912), relativity was interpreted as an antimetaphysical and instrumentalist tract and was hailed as "a mighty impulse for the development of the philosophical point of view of our time." When in London on 6 November 1919 the result of the British eclipse expedition was revealed to bear out one of the predictions of general relativity theory, the discussion of implications rose to fever pitch among scholars and laymen, beginning with declarations such as that in *The Times* of London (8 November 1919): the theory had served "to overthrow the certainty of ages, and to require a new philosophy, a philosophy that will sweep away nearly all that has hitherto been accepted as the axiomatic basis of physical thought." It became evident that, as Newton had "demanded the muse" after the *Principia*, now it would be Einstein's turn.

In fact, Einstein did his best to defuse the euphoria and excess of attention that engulfed and puzzled him from that time on. When asked to explain the mass enthusiasm, his answer in 1921 was that "it seemed psychopathological." The essence of the theory was chiefly "the logical simplicity with which it explained apparently conflicting facts in the operation of natural law," freeing science of the burden of "many general assumptions of a complicated nature."[4] That was all. As for being labeled a great revolutionary, as his friends and opponents did equally, Einstein took every opportunity to disavow it. He saw himself essentially as a continuist, had specific ideas on the way scientific theory developed by evolution,[5] and attempted to keep the discussion limited to work done and yet to be done in science. He did not get much help, however. Thus, the physicist J.J. Thomson reported that the Archbishop of Canterbury, Randall Davidson, had been told by Lord Haldane "that relativity was going to have a great effect upon theology, and that it was his duty as Head of the English Church to make himself acquainted with it....The Archbishop, who is the most conscientious of men, has procured several books on the subject and has been trying to read them, and they have driven him to what it is not too much to say is a state of *intellectual desperation*." On Einstein's first visit to England in June 1921, the Archbishop of Canterbury therefore sought him out to ask what effect relativity would have on religion. Einstein replied briefly and to the point: "None. Relativity is a purely scientific matter and has nothing to do with religion."[6] But of course this did not dispose of the question. Later that year, even the scientific journal *Nature* felt it necessary to print opposing articles on whether "Einstein's space-time is the death knell of materialism."[7]

Although the crest of the flood, and the worst excesses, have now passed, debates of this sort continue. More constructively, since philosophy is concerned in good part with the nature of space and time, causality, and other conceptions to which relativity and quantum physics have contributed, Einstein has had to be dealt with in the pages of philosophers, from Henri Bergson and A.N. Whitehead to the latest issues of the professional journals. As John Passmore observed, it appeared in this century that "physics fell heir to the responsibility of metaphysics."[8]

Some philosophers have confessed that Einstein's work started them off on their speculations in the first place, thus giving some direction to their very careers. One example is Karl Popper, who in his autobiography revealed that his falsification criterion originated in his interpretation of a passage in Einstein's popular exposition of relativity, which Popper says he read with profound effect when he was still in his teens.

Philosophy was no doubt destined to be the most obvious and often the earliest and most appropriate field, outside science itself, that the radiation from Einstein's work would reach. But soon there were others, as several essays in this collection show, even though the connections made or asserted were not always valid. From Einstein's wide-ranging output, relativity was invoked most frequently. Cultural anthropology, in Claude Lévi-Strauss's phrase, had evolved the doctrine of cultural relativism "out of a deep feeling of respect toward other cultures than our own;" but this doctrine often invited confusion with physical relativity. Much that has been written on "ethical relativity" and on "relativism" is based on a seductive play with words. And painters and art critics have helped to keep alive the rumor of a supposed genetic connection of visual arts with Einstein's 1905 publication.

Here again, Einstein protested when he could and, as so often, without effect. One art historian submitted to him a draft of an essay entitled "Cubism and the Theory of Relativity," which argued for such a connection—for example, that in both fields "attention was paid to relationships, and allowance was made for the simultaneity of several views."[9] Politely but firmly, Einstein tried to put him straight, and he explained the difference between physical relativity and vulgar relativism so succinctly as to invite an extensive quotation:

> The essence of the theory of relativity has been incorrectly understood in it [your paper], granted that this error is suggested by the attempts at popularization of the theory. For the description of a given state of facts one uses almost always only one system of coordinates. The theory says only that the general laws are such that their form does not depend on the choice of the system of coordinates. This logical demand, however, has nothing to do with how the single, specific case is represented. A multiplicity of systems of coordinates is not needed for its representation. It is completely sufficient to describe the whole mathematically in relation to one system of coordinates.
>
> This is quite different in the case of Picasso's painting, as I do not have to elaborate any further. Whether, in this case, the representation is felt as artistic unity depends, of course, upon the artistic antecedents of the viewer. This new artistic "language" has nothing in common with the Theory of Relativity.[10]

Einstein might well have added here, as he did elsewhere, that the existence of a multiplicity of frames, each one as good as the next for solving some problems in mechanics, went back to the seventeenth century (Galilean relativity). As to the superposition of different aspects of an object on a canvas, that had been done for a long time; thus Canaletto drew various parts of a set of buildings from different places and merged them in a combined view on the painting (for example, *Campo S. S. Giovanni e Paolo*), the view becoming thereby an impossible *veduta*.

It was therefore doubly wrong to invoke Einstein as authority in support of the widespread misunderstanding that physical relativity meant that all frameworks, points of view, narrators, fragments of plot or thematic elements are created equal, that each of the polyphonic reports and contrasting perceptions is as valid or expedient as any other, and that all of these, when piled together or juxtaposed, *Rashomon*-like, somehow constitute the real truth. If anything, twentieth-century relativistic physics has taught the contrary: that under certain conditions we can extract from different reports, or even from the report originating in one frame properly identified, all the laws of physics, each applicable in any framework, each having therefore an invariant meaning, one that does not depend on the accident of which frame one inhabits. It is for this reason that, by comparison with classical physics, modern relativity is simple, universal, and, one may even say, "absolute." The cliché became, erroneously, that "everything is relative," whereas the whole point is that out of the vast flux one can distill the very opposite: "some things are invariant."

The cost of the terminological confusion has been so great that a brief elaboration on this point will be relevant. Partly because he saw himself as a continuist rather than as an iconoclast, Einstein was reluctant to present this new work as a new *theory*. The term "relativity theory," which made the confusions in the long run more likely, was provided by Planck and Abraham in 1906. For a time Einstein referred to it in print as the "so-called relativity theory," and until 1911 he avoided using the term altogether in the titles of his papers on the subject. In his correspondence Einstein seemed happier with the term *Invariantentheorie*, which is of course much more true to its method and aim. How much nonsense we might have been spared if Einstein had adopted that term, even with all its shortcomings! To a correspondent who suggested just such a change, Einstein replied (letter to E. Zschimmer, 30 September 1921): "Now to the name relativity theory. I admit that it is unfortunate, and has given occasion to philosophical misunderstandings.... The description you proposed would perhaps be better; but I believe it would cause confusion to change the generally accepted name after all this time."

To come back to Einstein's careful disavowal of a substantive genetic link between modern art and relativity: far from abandoning the quest for it, his correspondent forged onward enthusiastically and published three such essays instead of one. Newton did not always fare better at the hands of eighteenth-century literati and divines who thought they were following in his footsteps. Poets rush in where scientists no longer fear to tread. And why not, if the apparent promises are so great? In April 1921, at the height of what Einstein on his first journey to the United States all too easily diagnosed as a pathological mass reaction, William Carlos Williams published a poem entitled "St. Francis Einstein of the Daffodils,"[11] containing such lines as "April Einstein / ...has come among the daffodils / shouting / that flowers and men / were created / relatively equal..." Declaring simply that "relativity applies to everything"[12] and that "relativity gives us the clue. So, again, mathematics comes to the rescue of the arts," Williams felt encouraged to adopt a new variable measure for his poems—calling it "a *relatively* stable foot, not a rigid one"[13]—that proved of considerable influence on other poets.

Williams was of course not alone. Robert Frost, Archibald MacLeish, e.e. cummings, Ezra Pound, T. S. Eliot, and some of their disciples (and outside the English-speaking world, others such as Thomas Mann and Hermann Broch) referred directly to Einstein or to his work. Some were repelled by the vision thought to be opened by the new science, but there were at least as many who seemed to be in sympathy with Jean-Paul Sartre's remark that "the theory of relativity applies in full to the universe of fiction."[14] Perhaps the most cheerful of the attempts to harness science and literature to common purpose is Lawrence Durrell's entertaining set of novels, *The Alexandria Quartet*, of which its author says by way of preface: "Modern literature offers us no Unities, so I have turned to science and am trying to complete a four-decker novel whose form is based on the relativity proposition. Three sides of space and one of time constitute the soupmix recipe of a continuum."[15] The intention is to use the properties of space and time as determining models for the structure of the book. Durrell says "the first three parts...are to be deployed spatially...and are not linked in a serial form....The fourth part alone will represent time and be a true sequel."

For that alone one would not have had to wait for Einstein. But more seems to be hoped for; that, and the level of understanding, is indicated by the sayings of the character Pursewarden in the novel. Pursewarden—meant to be one of the foremost writers in the English language, his death mask destined to be placed near those of Keats and Blake—is quoted as saying, "In the Space and Time marriage we have the greatest Boy meets Girl story of the age. To our great-grandchildren this will be as poetical a union as the ancient Greek marriage of Cupid and Psyche seems to us." Moreover, "the Relativity proposition was directly responsible for abstract painting, atonal music, and formless...literature."[16]

Throughout the novel it is evident that Durrell has taken the trouble to read up on relativity, but chiefly out of impressionistic popularizations such as *The Mysterious Universe* by James Jeans, even though Durrell readily confessed that "none of these attempts has been very success-ful."[17] There is something touching and, from the point of view of an in-tellectual historian, even a bit tragic about the attempt. In his study, *A Key to Modern British Poetry*, Durrell revealed his valid concern to show that as a result of "the far-reaching changes in man's ideas" about the outer and inner universe, "language has undergone a change in order to keep in line with cosmological inquiry (of which it forms a part)."[18] Yet on page after page the author demonstrates that he has been misled by the simplifi-cations of H. V. Routh and James Jeans; he believes that Rutherford and Soddy suggested that the "ultimate laws of nature were not simply causal at all," that "Einstein's theory joined up subject and object," that "sofar as phenomena are concerned...the uniformity of nature disappears," and so forth.[19] The terrible but clarifying remark of Wolfgang Pauli comes to mind, who said about a theory that seemed to him doomed: "It is not even wrong."

If I have spelled out some of the misunderstandings by which Ein-stein's work, for better or worse, has been thought to have found its way into contemporary culture, I had to do so in justice to any occasion associ-ated with his name. But the examples of incorrect interpretation prepare us to appreciate the correct ones that much more. I should confess that my own favorite example of the successful transmutation of scientifically based conceptions in the writer's imagination is a novel, and a controver-sial one. William Faulkner's *The Sound and the Fury* is more like an earthquake than a book. Immediately on publication in 1929 it caused uni-versal scandal; for example, not until Judge Curtis Bok's decision in 1949 was this, among Faulkner's others novels, allowed to be sold in Philadel-phia. On the surface it seems unlikely that this book—even a friendly re-viewer characterized it as "designedly a silo of compressed sin"—has any resonance with the ideas of modern physics, by intent or otherwise. At the time he poured himself into the book, Faulkner was still almost unknown, largely self-taught, eking out a meager living as a carpenter, hunter, and coal carrier on the night shift of a power station, his desk the upturned wheelbarrow on which he would write while kneeling on the floor. Yet even there he was not isolated if he read only a small part of the flood of articles in newspapers, periodicals, and popular books in the 1920s dealing with the heady concepts of relativity theory—such as the time dilation experienced by a clock traveling through space, the necessity of recognizing the meaninglessness of absolute time and space—and the recent quantum physics, with its denial of the comforts of classical causality. Particularly in America, Einstein was quoted down to the level of local evening papers and Popular Mechanics, resulting in wide circulation of such haunting epigrams as his remark, made in exasperation to Max Born (1926), that "God does not throw dice." Could any of this have reached Faulkner?

In the second of the four chapters of *The Sound and the Fury* we follow Quentin Compson of Jefferson, Mississippi, as he lives through a day in June 1910. It is the end of his freshman year at college and the culmination of a short life wrenched by the degeneration and guilt, the fixations and tribal racism of his whole haunted family—from his father Jason, drinking himself to death, to his idiot brother Benjamin, whose forty-acre pasture has been sold to send Quentin to college. The only resource of human affection he has known came from black laborers and servants, although they have been kept in the centuries-old state of terror, ignorance, and obeisance. But the Compsons are doomed. As the day unfolds, Quentin moves towards the suicide he knows he will commit at midnight.

It is all too easy to discover theological and Freudian motives woven into the text, and one must not without provocation drag an author for cross-examination into the physics laboratory when he has already suffered through interrogations at the altar and on the couch. But Faulkner asked for it. Let me select here from a much more extensive body of evidence in the novel itself.

Quentin's last day on earth is a struggle against the flow of time. He attempts to stem the flow, first by deliberately breaking the cover glass of the pocket watch passed down to him from grandfather and father, then twisting off the hands of the watch, and then launching on seemingly random travel, by streetcar and on foot, across the whole city. His odyssey brings him to the shop of an ominous, cyclopean watch repairer. Quentin forbids him to tell him the time but asks if any of the watches in the shop window "are right." The answer he gets is "No." But wherever he then turns, all day and into the night, he encounters chimes, bells ringing the quarter-hours, a factory whistle, a clock in the Unitarian steeple, the long, mournful sound of the train tracing its trajectory in space and time, "dying away, as though it were running through another month." Even his stomach is a kind of space-time metronome. "The business of eating inside of you space too space and time confused stomach says noon brain says eat o'clock All right I wonder what time it is what of it." Throughout, Quentin carries the blinded watch with him, the watch that never knew how to tell real time and cannot even tell relative time. But it is not dead: "I took out my watch and listened to it clicking away, not knowing it could not even lie."[20] And in the streetcar, the clicking away of time is audible to him only while the car has come to a stop.

Quentin has taken a physics course that freshman year and uses it to calculate how heavy the weights must be that he buys to help drown himself. It is, he says wryly to himself, "the only application of Harvard," and as he reflects on it: "The displacement of water is equal to the something of something. Reducto absurdum of all human experience, and two six-pound flat-irons weigh more than one tailor's goose. What a sinful waste, Dilsey would say. Benjy knew it when Damuddy died. He cried."

As midnight approaches, before he is ready to put his "hand on the light switch" for the last time,[21] Quentin is overcome by torment caused by the shamed memory of his incestuous love for this sister Candace, by her loss, and by his own sense of loss even of the meaningfulness of that double betrayal. In anguish he remembers his father's terrible prediction after he had made his confession:

> You cannot bear to think that some day it will no longer hurt you like this now were getting at it...you wont do it under these conditions it will be a gamble and the strange thing is that man who is conceived by accident and whose every breath is a fresh cast with dice already loaded against him will not face that final main which he knows beforehand he has assuredly to face without essaying expedients...that would not deceive a child until some day in very disgust he risks everything on a single blind turn of a card no man ever does that under the first fury of despair or remorse or bereavement he does it only when he has realized that even the despair or remorse or bereavement is not particularly important to the dark diceman...it is hard believing to think that a love or a sorrow is a bond purchased without design and which matures willynilly and is recalled without warning to be replaced by whatever issue the gods happen to be floating at the time.

This was not the God Newton had given to his time—Newton, of whom, just two centuries before Faulkner's soaring outcry, the poet James Thomson had sung in 1729 that "the heavens are all his own, from the wide role of whirling vortices, and circling spheres, to their great simplicity restored." Nor, of course, was it Einstein's God, a God whose laws of nature are both the testimony of His presence in the universe and the proof of its saving rationality. But this, it seems to me, defines the dilemma precisely. If the poet neither settles for the relief of half-understood analogies nor can advance to an honest understanding of the rational structure of that modern world picture, and if he is sufficiently sensitive to this impotency, he must rage against what is left him: time and space are then without meaning; so is the journey through them; so is grief itself, when the very gods are playing games of chance, and all the sound and the fury signify nothing. And this leads to recognizing the way out of the dilemma, at least for a few. At best, as in the case of Faulkner, this rage itself creates the energy needed for a grand fusion of the literary imagination with perhaps only dimly perceived scientific ideas. There are writers and artists of such inherent power that the ideas of science they may be using are dissolved, like all other externals, and rearranged in their own glowing alchemical cauldron.

It should not, after all, surprise us; it has always happened this way. Dante and Milton did not use the cosmological ideas of their time as tools to demarcate the allowed outline or content of their imaginative constructs. Those students of ours who, year after year, write dutifully more or less the same essay, explaining the structure of the *Divine Comedy* or *Paradise Lost* by means of astronomy, geography, and the theory of optical phenomena—they may get the small points right, but they miss the big one, which is that the good poet is a poet surely because he can transcend rather than triangulate. In Faulkner, in Eliot's *The Waste Land*, in Woolf's *The Waves*, in Mann's *Magic Mountain* it is futile to judge whether the traces of modern physics are good physics or bad, for these trace elements have been used in the making of a new alloy. It is one way of understanding Faulkner's remark on accepting his Nobel Prize in 1950: the task was "to make out of the material of the human spirit something which was not there before."[22] And insofar as an author *fails* to produce the feat of recrystallization, I suspect this lack would not be cured by more lessons on Minkowski's space-time, or Heisenberg's indeterminacy principle, or even thermodynamics, although these lessons could occasionally have a prophylactic effect that might not be without value.

Here we suddenly remember that, of course, the very same thing is true for scientists themselves. The most creative ones, almost by definition, do not build their constructs patiently by assembling blocks that have been precast by others and certified as sound. On the contrary, they too melt down the ready-made materials of science and recast them in a way that their contemporaries tend to think is outrageous. That is why Einstein's own work took so long to be appreciated even by his best fellow physicists, as I noted earlier. His physics looked to them like alchemy, not because they did not understand it at all, but because, in one sense, they understood it all too well. From their thematic perspective, Einstein's was anathema. Declaring, by simple postulation rather than by proof, Galilean relativity to be extended from mechanics to optics and all other branches of physics; dismissing the ether, the playground of most nineteenth-century physicists, in a preemptory half-sentence; depriving time intervals of inherent meaning; and other such outrages, all delivered in a casual, confident way in the first, short paper on relativity—those were violent and "illegitimate" distortions of science to almost every physicist. As for Einstein's new ideas on the quantum physics of light emission, Max Planck felt so embarrassed by it when he had to write Einstein a letter of recommendation seven years later that he asked that this work be overlooked in judging the otherwise promising young man.

Moreover, the process of transformation characterizes not only science itself and the flow of ideas from high science to high literature. It also works across the boundaries in other ways. The most obvious example is Einstein's importation into his early physics of an epistemology that he himself thought, with some enthusiasm, to be based on Ernst Mach's kind of positivism. Mach had begun to find him out on this point even while Einstein was still signing his letters to Mach as "Your devoted student."

It seems clear to me that without this process of transformation, willing or unwilling, of ideas from science and from philosophy, physics would not have come into its twentieth-century form. (A similar statement may well be made for the cases of Copernicus, Kepler, Galileo, and Newton.) The case of Einstein suggests, therefore, that the accomplishments of the major innovators—and not only in science—depend on the ability to persevere in four ways: by being loyal primarily to one's own belief system rather than to the current faith; by perceiving and exploiting the man-made nature and plasticity of human conceptions; by demonstrating eventually that the new unity that has been promised does become lucid and convincing to lesser mortals active in the same field—that he has it all "wrong" in the right way; and, in those rare cases, even by issuing ideas that lend themselves, quite apart from misuse and oversimplification, to further adaptation and transformation in the imagination of similarly exalted spirits who live on the other side of disciplinary boundaries.

It remains to deal with one more, somewhat different mechanism by which Einstein's imprint came to be felt far beyond his own field of primary attention: the power of his personal intervention on behalf of causes ranging from the establishment of a homeland for a persecuted people to his untiring efforts, over four decades, for peace and international security. In retrospect we can see that he had the skill, at strategic periods of history, to lend his ideas and prestige to the necessary work of the time. Even the most famous of these personal interventions, the call to President Roosevelt in 1939 to initiate a study on whether the laws of nature allow anyone to produce an atomic weapon, was of that sort, although it has perhaps been misunderstood more widely than anything else Einstein did. He was, after all, correct in his perception that the Germans, who were pushing the world into a war, had all the skill and intention needed to start production of such a weapon if it was feasible. In fact, they had a headstart, and but for some remarkable blunders and lack of means they might have fulfilled the justified fears, with incalculable consequences on the course of civilization.

To highlight these personal interactions, I select one as more or less paradigmatic of the considerable effect Einstein had even in some brief or seemingly casual discussion with the right person. The Swiss psychologist Jean Piaget's work entered its most important phase with the publication in 1946 of *The Child's Conception of Time*. The book begins with a plain acknowledgment: "This work was promoted by a number of questions kindly suggested by Albert Einstein more than fifteen years ago (1928, at a meeting in Davos).... Is our intuitive grasp of time primitive or derived? Is it identical with our intuitive grasp of velocity? What if any bearing do these questions have on the genesis and development of the child's conception of time? Every year since then we have made a point of looking into these questions.... The results (concerning time) are presented in this volume; those bearing on the child's conception of motion and speed are reserved for a later work."[23] Throughout his later writings, Piaget remarks on this debt: "It was the author of the theory of relativity who suggested to us our work,"[24] or "Einstein once suggested we study the question from the psychological viewpoint and try to discover if there existed an intuition of speed independent of time."[25] In addition, Piaget refers explicitly to notions of relativity and other aspects of Einstein's work.[26] Since Einstein came to have an immense correspondence with leaders in virtually every type of endeavor, it will not be difficult to reconstruct instances of the same sort as Einstein's brief but seminal interactions with Piaget.[27]

Looking back at the variety of ways in which Einstein came to impress the imagination of his time and ours, we can discern some rough categories, spread out, as it were, in a spectrum from left to right. At the center portion, corresponding to the largest intensity, one finds the widespread but unfocused and mostly uninformed fascination, manifested in a variety of ways, from enthusiastic mass gatherings to glimpse the man, to the outpouring of popularizations with good intentions, to responses that betray the vague discomfort aroused by the ideas. A good example of the last is an editorial entitled "A Mystic Universe" in *The New York Times* of 28 January 1928 (p. 14): "The new physics comes perilously close to proving what most of us cannot believe...Not even the old and much simpler Newtonian physics was comprehensible to the man on the street. To understand the new physics is apparently given only to the highest flight of mathematicians...We cannot grasp it by sequential thinking. We can only hope for dim enlightenment." The editorial writer then notes that the ever-changing scene in physics does offer some "comfort":

> Earnest people who have considered it their duty to keep abreast of science by readapting their lives to the new physics may now safely wait until the results of the new discoveries have been fully tested out by time, harmonized and sifted down to a formula that will hold for a fair term of years. It would be a pity to develop an electronic marriage morality and find that the universe is after all ether, or to develop a wave code for fathers and children only to have it turn out that the family is determined not by waves but by particles. Arduous enough is the task of trying to understand the new physics, but there is no harm in trying. Reshaping life in accordance with the new physics is no use at all. Much better to wait for the new physics to reshape our lives for us as the Newtonian science did.

Similarly, in Tom Stoppard's play *Jumpers* a philosopher is heard to ask: "If one can no longer believe that a twelve-inch rule is always a foot long, how can one be sure of relatively less certain propositions?"

Near this position, as we said, are the enthusiastic misapplications, usually achieved by an illicit shortcut of meaning from, say, the true statement that the operational definition of length is "framework"-dependent, to the invalid deduction that mental phenomena in a human observer have *thereby* been introduced into the very definitions of physical science. The irony here is that the first lessons we learned from relativity physics were that short circuits in signification must be avoided, for they were just what burdened down classical physics, and that attention must be paid as never before to the meanings of the terms we use.

When we now glance further toward the left, or blue, end of the spectrum, the expressions of resignation or futility become more explicit. Indeed, among some of the most serious intellectuals there seems, on this point, a sense of despair. By the very nature of their deep motivation they must feel most alienated from a universe whose scientific description they can hardly hope to understand except in a superficial way. The much-admired humanistic scholar Lionel Trilling spoke for many when he stated the dilemma frankly and eloquently:

> The operative conceptions [of science] are alien to the mass of educated persons. They generate no cosmic speculation, they do not engage emotion or challenge imagination. Our poets are indifferent to them...
>
> This exclusion of most of us from the mode of thought which is habitually said to be the characteristic achievement of the modern age is bound to be experienced as a wound given to our intellectual self-esteem. About this humiliation we all agree to be silent, but can we doubt that it has its consequences, that it introduced into the life of mind a significant element of du-

biety and alienation which must be taken into account in any estimate that is made of the present fortunes of mind?[28]

Einstein, who had intended originally to become a science teacher, came to understand this syndrome, and the obligation it put on him. He devoted a good deal of time to popularization of his own. His avowed aim was to simplify short of distortion. In addition to a large number of essays and lectures, he wrote, and repeatedly updated, a short book on relativity that he promised in the very title to be *gemeinverständlich*.[29] It is, however, overly condensed for the most nonscientific beginners. Later, Einstein collaborated with Leopold Infeld in a second attempt to reach out to the population at large by means of a book-length treatment of modern physics. As the preface acknowledged, the authors no longer attempted "a systematic course in elementary facts and theories." Rather, they aimed at a historical account of how the ideas of relativity and quanta entered science, "to give some idea of the eternal struggle of the inventive human mind for a fuller understanding of the laws governing physical phenomena."[30] In fact, there is to this day no generally agreed source, the *reading* of which by itself will bring a large fraction of its nonscientific audiences to a sound enough understanding of these ideas, even for those who truly want to attain it and are willing to pay close attention. I believe it is a fact of great consequence that it takes a much larger effort, and one starting earlier than most people undertake. To make matters worse, so little has been found out about how scientific literacy is achieved or resisted that little blame can be spun off on the would-be students, young or old.

Going now further along the spectrum in the same direction we encounter outright hostility and opposition to Einstein's work, either on scientific or on ideological grounds. Almost all scientists, even those initially quite reluctant, became eventually at least reconciled, save (to this day) for Einstein's famous refusal to regard the statistical interpretation as fundamental. On the other hand, the opposition to Einstein's work on grounds other than scientific has had a longer history. Thus, a number of studies now exist that show the lengths to which various totalitarian groups, for various reasons, felt compelled to go in their attacks.

Turning now to the other, more "positive" half of the spectrum, we see there the gradual acceptance and elaboration of Einstein's work within the corpus of physical science; its penetration into technology (largely unmarked) and into the more thoughtful philosophies of science; Einstein's effect through his personal intervention, causing some historic redirections of research; and its passage into the scientific world-picture of our time, as it tries to achieve a unification that eluded Einstein. And beyond that, at the end of the spectrum where the number of cases is small but the color deep and vibrant, we perceive the examples of creative transformation beyond science. Those are the works of the few who found that scientific ideas, or rather *metaphors* embodying such ideas, released in them a fruitful response with an authenticity of its own, far removed from textbook physics.

This last is the oldest and surely still the most puzzling interplay between science and the rest of culture. Evidently, the mediation occurs through a sharing of an analogy or metaphor—irresistible, despite the dangers inherent in the obvious differences or discontinuities. We know that such a process exists, because any major work of science itself, in its nascent phase, is connected analogically rather than fully logically, both with the historic past in that science and with its supporting data. The scientist's proposal may fit the facts of nature as a glove fits a hand, but the glove does not uniquely imply the hand, nor the hand the glove.

Einstein spoke insistently over the decades about the need to recognize the existence of such a discontinuity, one that in his early scientific papers asserted itself first in his audacious method of postulation. In essay after essay, he tried to make the same point, even though it had little effect on the then reigning positivism. Typical are the phrases in his Herbert Spencer Lecture of 1933.[31] The rational and empirical components of human knowledge stand in "eternal antithesis," for "propositions arrived at by purely logical means are completely empty as regards reality." In this sense, the "fundamentals of scientific theory," being initially free inventions of the human mind, are of "purely fictional character." The phenomenic-analytic dichotomy makes it inherently impossible to claim that the principles of a theory are "deduced from experience" by "abstraction," that is, by logically complete claims of argument.

If this holds for the creative act in the individual's pursuit of science itself, we should hardly be surprised to find the claim to be extended to more humanistic enterprises. The test, in both instances, is of course whether the freely invented coatcheck yields the intended use of a suitable overcoat. The existence of both splendid scientific theories and splendid products of the humanistic imagination shows that despite all their other differences, they share the ability to build on fundamentals of a "purely fictional character." And even the respective fundamentals, despite all their differences, can share a common origin. That is to say, at a given time the cultural pool contains a variety of themata and metaphors, and some of these have a plasticity and applicability far beyond their initial provenance. The innovator, whether a scientist or not, necessarily dips into this pool for his fundamental notions and in turn may occasionally deposit into it new or modified themata and metaphors of general power. Examples of such science-shaped metaphor, each of these by no means a "fact" of the external world, yet revealing immense explanatory energy, are easy to find: Newton's "innate force of matter (*vis insita*)" and the Newtonian clockwork universe; Faraday's space-filling electric and magnetic lines of force; Niels Bohr's examples of complementarity in physics and in daily life; Mendeleev's neat tableau setting for the families of elements and Rutherford's long parent-daughter-granddaughter chains of decaying atoms; Minkowski's space-time "World," of which our perceptible space and time are like shadows playing on the wall of Plato's cave; and of course the imaginative scenes Einstein referred to—the traveler along the light beam, the calm experimenter in the freely falling elevator, the dark, dice-playing God, the closed but unbounded cosmos, the Holy Grail of complete unification of all forces of nature. So it continues in science.

The allegorical use of such conceptions may, as we have noted, help to shape works of authenticity outside the sciences. And the process works both ways. Thus Niels Bohr acknowledged that his reading in Kierkegaard and William James helped him to the imaginative leap embodied in his physics, Einstein stressed the influence on his early scientific thinking of the philosophical tracts of that period, and Heisenberg noted the stimulus of Plato's *Timeaus*, read in his school years. No matter if such "extraneous" elements are eventually suppressed or forgotten, or even have to be overcome, at an early point they can encourage the mind's struggle.

We conclude, then, that in pursuing the evident and documentable cases of "impact" of one person or field on another, we have been led to a more mysterious fact, namely, the mutual adaptation and resonance of the innovative mind with portions of the total set of metaphors current at a given time. The philosopher José Ortega y Gasset was one of those who struggled with this idea. In 1921–1922, evidently caught up by the rise of the new physics, he began an essay on "The Historical Significance of the Theory of Einstein."[32] There he remarked quite correctly that the most relevant issue was not that

> the triumph of the theory will influence the spirit of mankind by imposing
> on it the adoption of a definite route...What is really interesting is the in-
> verse proposition: the spirit of man has set out, of its own accord, upon a
> definite route, and it has therefore been possible for the theory of relativity
> to be born and to triumph. The more subtle and technical ideas are, the
> more remote they seem from the ordinary preoccupations of men, the more

authentically they denote the profound variations produced in the historical mind of humanity.[33]

But to this day, attempts to go much beyond that point, it seems to me, have not been very successful. The tantalizing task remains for future scholars.

References

1. Daniel N. Lapedes, ed., *McGraw-Hill Dictionary of Scientific and Technical Terms*, 2nd ed., New York, McGraw-Hill, 1978, pp. 512–513. As another measure in the continuing, albeit sometimes only ritualistic, reference made in the ongoing research literature to Einstein's publications, Eugene Garfield has found that during the period 1961-1975 the serious scientific journals *in toto* carried no less than 40 million citations to previously published articles. Of these, 58 cited articles stand out by virtue of having been published before 1930 *and* cited over 100 times each; and among these 58 enduring classics, ranging from astronomy and physics to biomedicine and psychology, 4 are Einstein's. See E. Garfield, *Current Contents* 21 (1976), pp. 5–9.

2. In Harry Woolf, ed., *Some Strangeness in the Proportion: A Centennial Symposium to Celebrate the Achievements of Albert Einstein*, Reading, MA: Addison-Wesley Publishing Co., Inc., 1980, p. 108.

3. Quoted in Ernst Lechner, *Physikalische Weltbilder*, Leipzig, Theodore Thomas Verlag, (1912), p. 84.

4. Quoted in the *New York Times*, 3,4, and 5 April 1921.

5. For discussion and documentation, see G. Holton, "Einstein's Search for the '*Weltbild*,'" ch. 4 in G. Holton, *The Advancement of Science and its Burdens*, New York, Cambridge University Press, 1986.

6. Quoted from J. J. Thomson, *Reflections and Recollections*, London: G. Bell and Sons, Ltd., 1936, p. 431 (italics in original). See also Philipp Frank, *Einstein: His Life and Times*, New York: Alfred A. Knopf, 1947, p. 190. Frank's book is one of the good sources for documentation on the reception and rejection of Einstein's theories by various religions and philosophic and political systems.

7. These articles, and excerpts from some other publications dealing with the influence of Einstein's work, have been gathered in L. Pearce Williams, ed., *Relativity Theory: Its Origins and Impact on Modern Thought*, New York: John Wiley and Sons, 1968. It is a useful compendium, and I am indebted to it for a number of illustrations to be referred to below.

8. John Passmore, *A Hundred Years of Philosophy*, rev. ed., New York: Basic Books, 1966, p. 332.

9. As reported by Paul M. LaPorte, "Cubism and Relativity, with a Letter of Albert Einstein," *Art Journal* 25, no. 3 (1966), p. 246.

10. Ibid. See also C.H. Waddington, *Behind Appearances: A Study of the Relations between Painting and the Natural Sciences in this Century*, Edinburgh: Edinburgh University Press, 1969, Cambridge, Mass: MIT Press, 1970, pp. 9–39. At the Jerusalem Symposium, Professor Meyer Schapiro presented an extensive and devastating critique of the frequently proposed relation between modern physics and modern art.

11. *Contact* 4 (1923), p. 3. I am indebted to Carol Donley's draft paper, "Einstein, Too, Demands the Muse" for this lead and others in the following paragraphs.

12. *Selected Essays of William Carlos Williams*, New York, Random House, 1954, p. 283.

13. Ibid., p. 340.

14. J.-P. Sartre, "François Mauriac and Freedom," in *Literary and Philosophical Essays*, New

York, Criterion Books, 1955, p. 23.
15. Lawrence Durrell, *Balthazar*, New York, E.P. Dutton, 1958, Author's Note, p. 9.
16. Ibid., p. 142.
17. For a good review of details, to which I am indebted, see Alfred M. Bork, "Durrell and Relativity," *Centennial Review* 7 (1963), pp. 191–203.
18. L. Durrell, *A Key to Modern British Poetry*, Norman, University of Oklahoma Press, 1952, p. 48.
19. Ibid., pp. 25, 26, 29.
20. All quotations are from William Faulkner, *The Sound and the Fury*, London, Chatto and Windus, 1961, pp. 81–177. I thank Dr. J. M. Johnson for a draft copy of her interesting essay, "Albert Einstein and William Faulkner," and have profited from some passages even while differing with others.
21. The chapter is shot through with references to light, light rays, even to travel "down the long and lonely light rays."
22. In *Les Prix Nobel en 1950*, Stockholm, Imprimierie Royale, 1951, p. 71.
23. Jean Piaget, *The Child's Conception of Time*, New York, Ballantine Books, 1971, p. vii.
24. Jean Piaget, *Genetic Epistemology*, New York, Columbia University Press, 1970. p. 69; see also p. 7.
25. Jean Piaget, *Psychology and Epistemology*, New York, Grossman Publishers, 1971, p. 82; see also pp. 10, 110. A similar statement is to be found in Piaget's *Six Psychological Studies*, New York, Vintage Books, 1968, p. 85.
26. For example, Jean Piaget, with Bärbel Inhelder, *The Child's Conception of Space*, New York, W. W. Norton, 1967, pp. 232–233; *The Child's Conception of Time*, London, Routledge and Kegan Paul, 1969, pp. 305–306; *Biology and Knowledge*, Chicago, University of Chicago Press, 1971, pp. 308, 337, 341–342. I wish to express my thanks to Dr. Katherine Sopka for help in tracing these references.
27. All this is quite apart from the role of model or personal culture hero that Einstein played in the lives of a great many individuals whom he never met. To this day, his picture can be found in wide circulation and in the most unlikely places, from the tee shirt of a student in high school to the workbench of a cobbler in Shanghai.
28. Lionel Trilling, *Mind in the Modern World: The 1972 Jefferson Lecture in the Humanities*, New York, Viking Press, 1972, pp. 13–14.
29. Albert Einstein, *Über die spezielle und die allgemeine Relativitätstheorie, gemeinverständlich*, Braunschweig, Vieweg, 1917. It was often translated and to this day is perhaps his most widely known work.
30. Albert Einstein and Leopold Infeld, *The Evolution of Physics*, New York, Simon and Schuster, 1938.
31. Albert Einstein, *On the Method of Theoretical Physics*, Oxford, Clarendon Press, 1933.
32. First published in English in 1933, in his *The Modern Theme*, New York, W. W. Norton.
33. Ibid., pp. 135–136.

Einstein and the Goal of Science

The notion of scientific progress is undergoing a transformation at the present time. There is little consensus on whether it is possible even in principle to determine the direction and magnitude of a change in scientific understanding or to what degree the activities of scientists are cumulative. On the contrary, much of the recent philosophical literature claims that science merely staggers from one fashion, conversion, revolution, or incommensurable exemplar to the next in a kind of perpetual, senseless Brownian motion, without discernible direction or goal. Some recent attempts to ascertain "progressiveness" by means of "rational reconstruction" have been unsuccessful primarily because they have not dealt with specific historical cases of actual scientific development. As to the scientists themselves, they are doing very well without a self-conscious analysis of their efforts and results. If pressed hard, they would tend to associate themselves with the assertion of George Sarton, the first of the modern historians of science:

> *Definition*: Science is systematized positive knowledge or what has been taken as such at different ages and in different places.
>
> *Theorem*: The acquisition and systematization of positive knowledge are the only human activities which are truly cumulative and progressive.
>
> *Corollary*: The history of science is the only history which can illustrate the progress of mankind. In fact, progress has no definite and unquestionable meaning in fields other than the fields of science.[1]

Apart from tending to sympathize with this sentiment, scientists are quite aware that scientific activity may be divided roughly into two kinds: that which is directed toward *analysis* and *accretion*, and that which is concerned with *synthesis*. For work of the former kind—by far the larger fraction—Pierre Duhem's description applies:

Physics makes progress because experiment constantly causes new disagreements to break out between laws and facts, and because physicists
constantly touch up and modify laws in order that they may more faithfully
represent the facts.[2]

The second type of scientific activity, synthesis, is rarer but more trans-
formative. Here progress is equivalent to an increase in inclusiveness (a
wider range of phenomena is accounted for by the new synthesis) or an
increase in parsimony or restrictiveness (fewer separate fundamental
terms and assumptions are needed). An exemplar is the heliocentric system of Copernicus. By virtue of its combination of inclusiveness and parsimony, Copernicus' system of planets had, as he explained, the great
merit that "not only [do the] phenomena follow therefrom, but also the order and size of all the planets and spheres and heaven itself are so linked
together that in no portion of it can anything be shifted without disrupting
the remaining parts and the universe as a whole." In his system, nothing is
arbitrary; there is no room for the ad hoc rearrangement of any orbit, as
had been possible before his work. Hence there is revealed a sparse ra-
tionale, a necessity that binds each detail to the whole design. Thereby it
carries the conviction that we understand why the planets are disposed as
they are and not otherwise.

The Galilean synthesis of celestial and terrestrial physics is another example, as is Newton's *Principia*, particularly in its theory of lunar motions. Newton's gravitational analysis allowed the prediction, from fundamentals, of the periods and magnitudes of known "inequalities." His
success put forward a vision that laws incorporating such concepts as
force and mass and the use of the proper mathematical and observational
techniques would suffice to discover an overwhelming unity of physical
science (one which Newton privately thought would eventually encompass the sphere of moral philosophy).

In our century, the incorporation of Newtonian mechanics into the general theory of relativity was a further example of scientific progress by
synthesis. Similarly, at the end of his *Principles of Quantum Mechanics*,
P. A. M. Dirac confidently declared that since "quantum mechanics may
be defined as the application of equations of motion to atomic particles,"
the domain of applicability of that theory "includes most of low-energy
physics and chemistry." These, he implied were now essentially solved
problems, the past rather than the future of science. However, the latter
beckoned:

> Now there are other kinds of interactions, which are revealed in high-en-
> ergy physics and are important for the description of atomic nuclei. These

interactions are not at present sufficiently well understood to be incorporated into a system of equations of motion....It is to be hoped that with increasing knowledge a way will be found for adapting the high-energy theories into a scheme based on equations of motion, and so unifying them with
those of low-energy physics.[3]

In almost every scientific research journal, this model of scientific progress by means of greater inclusiveness and parsimony can be found. For
example, in an article in *Science*, Victor Weisskopf chose six physical
constants—the mass of the proton, the mass of the electron and its electric
charge, the velocity of light, Newton's gravitational constant, and the
quantum of action of Planck—and a few fundamental laws (for example,
de Broglie's relations connecting momentum and energy with wavelength
and frequency, and the Pauli principle). Putting these data and laws together, he *derived* from them a host of predictions that in fact correspond
to facts of observation: the size and energy of nuclei, the mass and hard-
ness of solids such as rocks, the height of mountains, and the size of our
sun and similar stars.[4] All these seemed at first unconnected, as they are
spread over an enormous range of magnitudes. Now they are seen to be
different consequences of a small number of fundamental posits.

This is indeed Newton's program, the search for omniscience triumphant. And if any one of the six physical constants on Weisskopf's list,
now considered a separate entity, could be derived from the other five—
for example, if the charge of the electron were found to be deducible from
some of the other fundamental constants—that would mark further progress, indeed a great step forward.

A by-product of scientific progress achieved through the continual decrease in the number of fundamental axioms and, simultaneously, through
the increase in the range of phenomena covered is an emerging synergism
with respect to techniques and methods. Results in one field become useful in another, far distant one, and often in a surprising way. The theoretical tools of condensed-matter physics that handle macroscopic phenomena such as ferromagnetism, superconductivity, and superfluidity are now
understood in terms of concepts and theories applicable at the submicro-
scopic level, where they connect with problems of the structure of stellar
bodies, nuclear physics, particle physics, and field theories. Robert Sachs
wrote:

> Symmetry principles, and the concepts of gauge theories and spontane-
> ously broken symmetries, turn out to play a common role so that theorists
> working in one such field may borrow methods from another. Just as all
> fields have shared in common methods of solution of linear equations, it
> now appears that new approaches to the solution of non-linear equations,

discovered first in connection with hydrodynamic problems, may be of great importance for the understanding of both condensed matter and elementary particles.[5]

So much for the largely intuitive, "working-model" notion of scientific progress current among scientists. Evidently it has been adequate in the present surge of excellent work. Nonetheless, these and more explicit conceptions of scientific progress have come under severe attack. I have referred to this debate earlier, and briefly note here objections from some philosophers of science. They hold that the outcome of scientific activity is *by definition* inevitably progress, hence that the perception of progress is self-fulfilling and, therefore, illusory. Other philosophers of science claim that scientific work must be measured against universal criteria, drawn up as "statute law," so that "degenerating" programs can be identified by these philosophers and thus denied research support and publication. By these criteria, however, Boltzmann's kinetic-atomic research program would have been considered in a state of "degeneration" between 1880 and 1905. As one commentator correctly observed, "Fortunately, the scientific community at that time was more tolerant of dissent...Boltzmann, Lorentz, Smoluchowski, Gibbs, Jeans, and other followers of the kinetic program, including Planck and Einstein after 1900, did not have too much difficulty in getting their work published. But [today], with intense competition for research funds and journal space, application of the methodology of scientific research programs might well be disastrous....Beware of the philosopher kings and philosopher research administrators!"[6]

The Two-Dimensional Model

Instead of attempting to amend the largely intuitive models current among most scientists, or the predominantly ahistorical models among scholars outside science, I shall turn to the more useful task of examining a specific case. The case I have in mind will allow us to obtain a more disciplined model of scientific progress from the actual practice of a scientist who was knowledgeable about and sensitive to the philosophical underpinnings of modern science. For this purpose there seems to be no better point of departure than the credo of a scientist whose work to a large degree has determined the predominant direction of scientific advance in his field from the early part of this century to our day and whose approach has been absorbed, consciously or not, by most current practitioners. I mean of course Albert Einstein. He wrote extensively on the methods and direction of scientific development, and an appropriate entry for formulating a theory of scientific advance is his essay of 1933, "Zur Methodik der theoretischen Physik."[7] Philipp Frank, Einstein's biographer and colleague, called this the "finest formulation of his views on the nature of a physical theory."[8]

Einstein begins with an important warning, but one which, by itself, would seem to overturn the possibility that scientific progress could be achieved on any model: he notes first of all to pay "special attention to the relation between the content of a theory," on the one hand, and "the totality of empirical facts," on the other. These constitute the two "components of our knowledge," the "rational" and the "empirical"; these two components are "inseparable"; but they stand also, Einstein warns, in "external antithesis." To support this conception, Einstein refers to a dichotomy built into Western science. The Greek philosopher-scientists provided a lasting reason for having confidence in the achievement of the human intellect by introducing into Western thought the "miracle of the logical system." Logic, as in Euclid's geometry, "proceeds from step to step with such precision that every single one of its propositions was absolutely indubitable." But "propositions arrived at by purely logical means are completely empty as regards reality"; "through purely logical thinking we can attain no knowledge whatsoever of the empirical world." Einstein tells us that it required the seventeenth-century scientists to show that scientific knowledge "starts from experience and ends with it."

Up to this point, therefore, we are left with a thoroughly dualistic method for doing science. On the one hand, Einstein says, "the structure of the system is the work of reason"; on the other hand, "the empirical contents and their mutual relations must find their representation in the conclusions of the theory." Indeed, virtually all of Einstein's commentators have followed him in stressing this dualism—and have left it at that. It is a view of science of which there are many variants.

I consider it a two-dimensional view. It can be defended up to a point and may be summarized as follows. Science deals with two types of meaningful statements, namely, propositions concerning empirical matters that ultimately boil down to meter readings and other public phenomena, and propositions concerning logic and mathematics that ultimately boil down to tautologies. The first of these, the propositions concerning empirical matters of fact, can in principle be rendered in protocol sentences in ordinary language that command the general assent of the scientific community. I call these the phenomenic propositions. The second type of propositions, meaningful insofar as they are consistent within the system of accepted axioms, I call analytic propositions. As a mnemonic device, and also to do justice to Einstein's warning about the "eternally antithetical" nature of these propositions, one may imagine them as arranged along one or the other of two orthogonal axes defining the two dimensions of a plane within which scientific discourse usually takes place. A scientific statement, in this view, is analogous to an element of area in the plane, and the projections of the statement onto the two axes are the aspects of the statement that can be rendered, respectively, as a protocol of observation (for example, "the needle swings to the left") and as a protocol of calculation (for example, "use vector calculus, not scalars").

Now it has been the claim of most modern philosophies of science that trace their roots to empiricism or positivism that any scientific statement has "meaning" only insofar as it can be shown to have phenomenic and/or analytic components in this plane. And, indeed, in the past, this Procrustean criterion has amputated from science notions such as innate properties, occult principles, and all kinds of tantalizing questions for which the consensual mechanism could not produce sufficiently satisfying answers. A good argument can be made that the silent but general agreement to keep the discourse consciously in the phenomenic-analytic plane, where statements and routines can be shared, is the main reason science has been able to grow so rapidly in modern times. While the degree of consensus at the developing edge of science is usually far less than most model makers of science realize, consensus about an area arises rapidly after the edge has moved on. Hence, the two-dimensional model is widely used to characterize science when it is rewritten for pedagogic purposes or "rationalized" for epistemological discussion.

Limitations of the Two-Dimensional Model

Nevertheless, this two-dimensional view has its costs. It overlooks or denies the existence of active mechanisms at work in the day-to-day experience of those who are actually engaged in the pursuit of science; and it is of little help in handling questions every historian of science has to face consciously, even if the working scientist, happily, does not. To illustrate, let me mention two such problems. Both have to do with the direction of scientific advance, and both will seem more amenable to solution once the dualistic view is modified.

First, if sound scientific discourse is directed entirely by the dictates of logic and of empirical findings, why is science not one great totalitarian engine, taking everyone relentlessly to the same inevitable goal? The laws of reason, the perception or detection of phenomena, and human skills to deal with both are presumably distributed equally over the globe; and yet the story of, say, the reception of Einstein's theories is strikingly different in Germany and England, in France and the United States. On the level of *personal* choice or a research topic, why were some of Einstein's contemporaries so fatally attracted to ether-drift experiments whereas he himself, as he put it to his friend W. J. de Haas, thought it as silly and doomed to failure as trying to study dreams in order to prove the existence of ghosts? As to skills for navigating in the two-dimensional plane, Einstein and Bohr were rather well matched, as were Schrödinger and Heisenberg. And yet there were fundamental antagonisms in terms of programs, tastes, and beliefs, with occasional passionate outbursts among scientific opponents.

Or, again, how to understand the great variety of different personal styles of scientists, all engaged in what they agree to be the "same" problem? If science *were* two-dimensional, the work of scientists in a given field might sooner or later be governed by a rigid, uniformly accepted exemplar. The documented existence of pluralism at all times points to a fatal flaw in the two-dimensional model—and also to its cure.

A second question that escapes the simple model, and to which I have devoted a number of case studies over the years, is this: why are many scientists, particularly in the nascent phase of their work, willing to hold firmly, and sometimes at great risk, to what can only be called a suspension of disbelief about the possibility of falsification? Moreover, why do they sometimes do so without having any empirical evidence on their side, or even in the face of disconfirming evidence?

Among countless examples of this sort, Max Planck, responsible for the idea of the quantum but one of the most outspoken opponents to its corpuscular implications, cried out as late as 1927: "Must we really ascribe to the light quanta a physical reality?"—this four years after the publication and verification of Arthur H. Compton's findings. On the other hand, when it came to explaining the electron in terms of what Planck called "vibrations of a standing wave in a continuous medium," along the lines proposed by de Broglie and Schrödinger, Planck gladly accepted the idea and added that "these principles have already [been] established on a solid foundation"—all that before Planck had heard of any experimental evidence along the lines provided by Davisson and Germer.[9]

Einstein was even more daring. To select just one among many illustrations, in 1916, when he wrote his book *Über die spezielle und die allgemeine Relativätstheorie*, he had to acknowledge that his general theory of relativity so far had only one observable consequence, the precession of the orbit of Mercury, whereas the predicted bending of light and the red shift of spectral lines owing to the gravitational potential were too small to be then observed. Nevertheless, Einstein readily proclaimed, in a daring sentence with which he ended his book in its first fifteen printings, from 1917 through 1919: "I do not doubt at all that these consequences of the theory will also find their confirmation." It is an example of the suspension of disbelief, an important mechanism in the practice of experimental and theoretical scientists.[10]

The Role of Presuppositions

What, then, must one conclude from Planck's predisposition for the continuum and against discreteness, from Einstein's predisposition for a theory that encompasses a wide rather than a narrow range of phenomena and so allows him to risk his reputation on a daring prediction, and from many other examples of suspension of disbelief in the face of missing tests and even (in the case of R.A. Millikan's oil-drop experiment) contrary data? Such cases serve to indicate that some *third mechanism* is present in determining the choices scientists make in the nascent phase of their work, in addition to the phenomenic and analytical mechanisms. Indeed, we can find it in Einstein's lecture on the method of theoretical physics: the two-dimensional model initially prominent in it gives way, on closer examination, to a more sophisticated and appropriate one. In addition to the two inseparable but antithetical components, there is indeed a third one, directed away from the plane bounded by the empirical and logical dimensions of the theory.

Einstein launches his argument by reminding his audience, as he often did, that the previously mentioned phenomenic-analytic dichotomy prevents the principles of a theory from being "deduced from experience" by "abstraction"—that is to say, by logical means. "In the logical sense [the fundamental concepts and postulates of physics are] free inventions of the human mind" and in that sense different from the unalterable Kantian categories. He repeats more than once that the "fundamentals of scientific theory" are of "purely fictitious character."[11] Or as he put it soon afterwards: the relationship between sense experience and concept "is analogous not to that of soup to beef, but rather to that of check number to overcoat."[12]

The essential arbitrariness of reference, Einstein explains, "is perfectly evident from the fact that one can point to two essentially different foundations"—the general theory of relativity and Newtonian physics—"both of which correspond with experience to a large extent," namely, with much of mechanics. Moreover, the elementary experiences do not provide a logical bridge to the basic concepts and postulates of either relativity or Newtonian mechanics. Rather, "the axiomatic basis of theoretical physics...must be freely invented."

With this declaration, Einstein has, of course, exposed the emptiness of all attempts to impose external standards of correct thinking on the practice of scientists or to condemn as "irrational" scientific work that fails to meet such criteria. Thus, he shows that good scientific reasoning follows the precepts of neither the Dionysians nor the Apollonians. Einstein is, however, quite aware that his insight leads immediately to a basic problem, and he spells it out: How "can we ever hope to find the right way? Nay, more, has this right way an existence outside our illusions? Can we hope to be guided safely by experience at all when there exist theories such as classical mechanics, which do justice to experience to a large extent, but without grasping the matter in a fundamental way?"

We have now left the earlier, confident portion of Einstein's lecture far behind. But at this very point, Einstein issues a clarion call: "I answer with full confidence that there is, in my opinion, a right way, and that we are capable of finding it." Here, Einstein goes suddenly beyond his earlier categories of empirical and logical efficacy and offers us a whole set of selection rules with which, as with a good map and compass, that "right way" to an implicit goal may be found. Here, there, everywhere, guiding concepts emerge and beckon from above the previously defined plane to point us on the right path.

The first directing principle Einstein mentions is his belief in the efficacy of formal structures. The "creative principle resides in mathematics"—not, for example, in mechanical models. Next there unfolds a veritable hymn to the guiding concept of simplicity. Einstein calls it "the principle of searching for the mathematically simplest concepts and their connections," and he cheers us on our way with many examples of how effective it has already proven to be:

> If I assume a Riemannian metric [in the four-dimensional continuum] and ask what are the *simplest* laws which such a metric can satisfy, I arrive at the relativistic theory of gravitation in empty space. If in that moment I assume a vector field or antisymmetrical tensor field which can be derived from it, and ask what are the simplest laws which such a field can satisfy, I arrive at Maxwell's equations for empty space.

And so on, collecting victories everywhere under the banner of simplicity.

Later in the lecture, we find two other guiding notions in tight embrace: the concept of parsimony or economy and that of unification. As science progresses, Einstein tells us, "the logical edifice" is more and more "unified," the "smaller [is] the number of logically independent conceptual elements which are necessary to support the whole structure." Higher up on that same page, we encounter nothing less than what he calls "the noblest aim of all theory," which is "to make these irreducible elements as simple and as few in number as is possible, without having to renounce the adequate representation of any empirical content."

Yet another guiding concept to which Einstein gladly confesses is the *continuum*, the field. From 1905 on, when the introduction of discontinuity in the form of light quantum forced itself on Einstein as a "heuristic" and therefore not fundamental point of view, he clung to the hope and program to keep the continuum as a fundamental conception, and he defended it with enthusiasm in his correspondence. It was part of what he called his "Maxwellian program" to fashion a unified field theory. Atomistic discreteness and all it entails was not the solution but rather the problem. So here he again considers the conception of "the atomic structure of matter and energy" to be "the great stumbling block for a unified field theory."

One cannot, he thought, settle for a basic duality in nature, giving equal status both to the field and to its antithesis. To be sure, neither logic nor experience forbade it. Yet it was almost unthinkable. As he wrote to his old friend, Michele Besso, "I consider it quite possible that physics might not, finally, be founded on the concept of field—that is to say, on continuous elements. But then out of my whole castle in the air—including the theory of gravitation, but also most of current physics—there would remain *nothing*."[13]

We have by no means come to the end of the list of presuppositions that guided Einstein. But it is worth pausing to note how plainly he seemed to have been aware of their operation in his own scientific work. In this too he was rare. Sir Isaiah Berlin, in his book *Concepts and Categories*, remarked: "The first step to the understanding of men is the bringing to consciousness of the model or models that dominate and penetrate their thought and action. Like all attempts to make men aware of the categories in which they think, it is a difficult and sometimes painful activity, likely to produce deeply disquieting results."[14] This is generally true; but it was not for Einstein, for at least two reasons. It was he, after all, who first realized the "arbitrary character" of what had for so long been accepted as "the axiom of the absolute character of time, viz., of simultaneity [which] unrecognizedly was anchored in the unconscious," as he put it in his "Autobiographical Notes." "Clearly to recognize this axiom and its arbitrary character really implies already the solution of the problem."[15] Giving up an explicitly or implicitly held presupposition has indeed often had the characteristic of the great sacrificial act of modern science. We find in the writings of Kepler, Planck, Bohr, and Heisenberg that such an act climaxes a period that in retrospect is characterized by the word "despair."

Having recognized and overcome the negative, or enslaving, role of presuppositions, Einstein also saw their positive, emancipating potential. In one of this earliest essays on epistemology, he wrote:

> A quick look at the actual development teaches us that the great steps forward in scientific knowledge originated only to a small degree in this [inductive] manner. For if the researcher went about his work without any preconceived opinion, how should he be able to at all select out those facts from the immense abundance of the most complex experience, and just those which are simple enough to permit lawful connections to become evident?[16]

Much later, in his "Reply to Criticisms," he reverted to the "eternal antithesis" by way of acknowledging that the distinction between "sense impressions," on the one hand, and "mere ideas," on the other, is a basic conceptual tool for which he can adduce no convincing evidence. Yet he needs this distinction. His solution is simply to announce: "We regard the distinction as a category which we use in order that we might the better find our way in the world of immediate sensation." As with other conceptual distinctions for which "there is also no logical-philosophical justification," one has to accept it as "the presupposition of every kind of physical thinking," mindful that "the only justification lies in its usefulness. We are here concerned with 'categories' or schemes of thought, the selection of which is, in principle, entirely open to us and whose qualification can only be judged by the degree to which its use contributes to making the totality of the contents of consciousness 'intelligible.'" Finally, he curtly dismisses an implied attack on these "categories" or "free conventions" with the remark: "Thinking without the positing of categories and concepts in general would be as impossible as is breathing in a vacuum."[17]

On Themata

Einstein's remarkable self-consciousness concerning his fundamental presuppositions throughout his scientific and epistemological writings allows one to assemble a list of about ten chief presuppositions—some of which

I have already noted—underlying his theory of construction throughout his long scientific career: primacy of formal (rather than materialistic or mechanistic) explanation; unity or unification; cosmological scale in the applicability of laws; logical parsimony and necessity; symmetry (as long as possible); simplicity; causality (in essentially the Newtonian sense); completeness and exhaustiveness; continuum; and, of course, constancy and invariance.

These ideals, to which Einstein was obstinately devoted, explain why he would continue his work in a given direction even when tests against experience were difficult or unavailable, or, conversely, why he refused to accept theories well supported by the phenomena but, as in the case of Bohr's quantum mechanics, based on presuppositions opposite to his own. Much of the same can be said of most of the major scientists whom I have studied, from Johannes Kepler to our contemporaries. Each has his own, sometimes idiosyncratic "map" of fundamental guiding notions that may be considered in principle separate, like the band structure of chromosomes, seen microscopically during mitosis of a cell nucleus.

With this finding, we must now reexamine our mnemonic device of the two-dimensional plane. I remove its insufficiency by defining a third axis, rising perpendicularly out of it. This is the dimension orthogonal to and not resolvable onto the phenomenic or analytic axes. Along it are located those fundamental presuppositions, often stable, many widely shared, that show up in the motivation of the scientist's actual work, as well as in the end product for which he strives. Decisions between them, insofar as they are consciously made, are judgmental (rather than, as in the phenomenic-analytic plane, capable in principle of algorithmic decidability). Since these fundamental presuppositions are not directly derivable either from observation or from analytic ratiocination, they require a term of their own. I call them *themata* (singular *thema*, from the Greek θέμα: that which is laid down, proposition, primary word).

On this view—and again purely as a mnemonic device—a scientific statement is no longer, as it were, an element of area on the two-dimensional plane but a volume-element, an entity in three-dimensional space, with components along each of the three orthogonal (phenomenic, analytic, and thematic) axes. The projection of the entity onto the two-dimensional plane continues to have the useful roles I stressed earlier; but for our analysis it is also necessary to consider the line element projected onto the third axis, the dimension of which one may imagine the range of themata to be entered. The statements of differing scientists are therefore like two volume-elements that do not completely overlap and so have some differences in their projections.

Scientists are generally not, and need not be, conscious of the themata they use (even if they are sometimes vaguely aware of them, as when they notice that the concepts they work with are "theory-laden"). But the historian of science can chart the growth of a given thema in the work of an individual scientist over time and show its power upon the scientist's imagination. Thematic analysis, then, is in the first instance the identification of the particular map of the various themata which, like fingerprints, can characterize an individual scientist, or a part of the scientific community, at a given time.

Most themata are ancient and long-lived. Many come in opposing dyads or triads that show up most strikingly during a conflict between individuals or groups that base their work on opposing themata. I have been impressed by the small number of thematic couples, or triads. A relative few have sufficed us throughout the history of the physical sciences. Thematic analysis of the same sort has begun to be brought to bear on significant cases in other fields.[18]

With this conceptual tool we can return to some of the puzzles we mentioned earlier. Where does the conceptual and even emotional support come from which, for better or worse, stabilizes the individual scientist's risky speculation and confident suspensions of disbelief during the nascent phase? The result of case studies is that choices and decisions of this sort are often made on the basis of loyal dedication to thematic presuppositions. Or again, if, as Einstein claimed, the principles are indeed free inventions of the human mind, should those not be an infinite set of possible axiom systems to which one could leap or cleave? Virtually every one of these would ordinarily be useless for constructing theories. How then could there be any hope of success, except by chance? The answer must be that the license implied in the leap to an axiom system of theoretical physics by the freely inventing mind is the freedom to make such a leap but not the freedom to make *any leap whatever*. The freedom is narrowly circumscribed by a scientist's particular set of themata that constrains and shapes the style, direction, and rate of advance of the engagement on novel ground.

And insofar as the individual sets of themata overlap, the progress of the scientific community as a group is similarly constrained or directed. Otherwise, the inherently anarchic connotations of "freedom" could indeed disperse the total effort. As Mendeleev wrote: "Since the scientific world view changes drastically not only from one period to another but also from one person to another, it is an expression of creativity....Each scientist endeavors to translate the world view of the school he belongs to into an indisputable principle of science." In practice, however, there is more coherence than this statement implies, and we shall presently look more closely at the mechanism responsible for it.

The "Need to Generalize"

Of all the problems that invite attention with these tools, the most fruitful example will be a return visit to that mysterious place, early in the 1933 essay, where Einstein speaks of the need to pay "special attention to the relations between the content of the theory and the totality of empirical fact." The *totality* of empirical fact! It is a phrase that recurs in his writings and indicates the sweep of his conscious ambition. But it does even more: it lays bare the most daring of all the themata of science and points to the holistic drive behind "scientific progress."

Einstein explicitly and frankly hoped for a theory that would ultimately be utterly comprehensive and completely unified. This vision drove him on from the special to the general theory and then to the unified field theory. The search for one grand, architectonic structure itself was of course not Einstein's invention. On the contrary, it is an ancient dream. At its worst, it has sometimes produced authoritarian visions that are as empty in science as their equivalent is dangerous in politics. At its best, it has propelled the drive to the various grand syntheses that rise above the more monotonous landscape of analytic science. This has certainly been the case in recent decades in the physical sciences. Today's defenders of the promise as applied to particle physics, who in the titles of their publications use the term "Grand Unification," are in a real sense the hopeful children of those earliest synthesis-seekers of physical phenomena, the Ionian philosophers.

To be sure, as Isaiah Berlin warned in *Concepts and Categories*, the quest for synthesis may be a trap. Berlin has christened it the "Ionian Fallacy," defined as the search, from Aristotle to Bertrand Russell to our own day, for the ultimate constituents of the world in some nonempirical sense. Superficially, the seekers of a unified physics, particularly in their monistic exhortations, appear to have fallen into that trap—from Copernicus, who confessed that the chief point of his work was to perceive nothing less than "the form of the world and the certain commensurability of its parts," to Max Planck, who exclaimed in 1915 that "physical research cannot rest so long as mechanics and electrodynamics have not been welded together with thermodynamics and heat radiation,"[19] to today's theorists who seem to follow the founding father of science among the ancient Greeks, Thales himself, in their insistence that one entity will explain all.

A chief point in my view of science is that scientists, insofar as they are successful as a profession, are in practice rescued from the trap that devotion to a single thema might eventually lead to by the actual variety of their thematic commitments. The multiplicity of views gives them as a group the flexibility that an authoritarian research program built on a single thema lacks. (There is an analogy here with the benefits of maintaining biodiversity among crop plants.) Something like an Ionian Enchantment, the commitment to the theme of grand unification, was upon Einstein. Once alerted, we can find it in his work from the very beginning, in his first published paper (1901), where he tried to understand the contrary-appearing forces of capillarity and gravitation. As he exclaims in a letter to his friend Marcel Grossmann, "It is a magnificent feeling to recognize the unity [*Einheitlichkeit*] of a complex of phenomena that to direct observation appear to be quite separate things"—in this case capillarity and gravitation, the physics of micro- and macro-regions. In each of his next papers we find something of the same drive, which he later called "my need to generalize." He examines whether the laws of mechanics provide a sufficient foundation for the general theory of heat and whether the fluctuation phenomena that turn up in statistical mechanics also explain the basic behavior of light beams and their interference, the Brownian motion of microscopic particles in fluids, and even the fluctuation of electric charges in conductors. And in his deepest work of those early years, in special relativity theory, the most powerful propellant is Einstein's drive toward unification. His clear motivation is to find a more general point of view that would subsume the seemingly limited and contrary problems and methods of mechanics and of electrodynamics. In the process he showed that electric and magnetic fields are aspects of one commonality viewed from different reference frames; that space and time are not separate; that energy and mass are fused in one conservation law; and, soon after, that reference systems with gravitation and with acceleration are equivalent. Again and again, previously separate notions were shown to be connected.

Following the same program obstinately to the end of his life, he tried to bring together, as he had put it once, "the gravitational field and the electromagnetic field into a unified edifice," leaving "the whole physics" as a "closed system of thought."[20] In his longing for a unified world picture—a structure that would yield deductively, and thus master, the "totality" of empirical facts—one cannot help hearing an echo of Goethe's Faust, who exclaimed that he longed "to detect the inmost force that binds the world and guides its course." For that matter, one hears Newton himself, who wanted to build a unifying structure so tight that the most minute details would not escape it.

The Unified Weltbild as "Supreme Task"

In its modern form, the Ionian Enchantment, expressing itself in the search for a unifying world picture, is usually traced to Alexander von Humboldt and Schleiermacher, Fichte and Schelling. The influence of the "Nature philosophers" on physicists such as Hans Christian Oersted— who was led by their ideas directly to the first experimental demonstration of the unification of electricity and magnetism—has been amply chronicled. At the end of the nineteenth century, in the Germany of Einstein's youth, the pursuit of a unified world picture as the scientist's highest task had become almost a cult activity. Looking on from the other side of the channel, J. T. Merz exclaimed in 1904 that the lives of the continental thinkers are

> devoted to the realization of some great ideal....The English man of science would reply that it is unsafe to trust exclusively to the guidance of a pure idea, that the ideality of German research has frequently been identical with unreality, that in no country has so much time and power been frittered away in following phantoms, and in systematizing empty notions, as in the Land of the Idea.[21]

Einstein himself could not easily have escaped notice of these drives toward unification even as a young person. For example, we know that as boy he was given Ludwig Büchner's widely popular book *Kraft und Stoff [Energy and Matter]*, a work Einstein recollected having read with great interest. The little volume does talk about energy and matter; but chiefly it is a late-Enlightenment polemic. Büchner comes out explicitly and enthusiastically in favor of an empirical, almost Lucretian scientific materialism, which Büchner calls a "materialistic world view." Through this world view, he declares, one can attain "the unity of energy and matter, and thereby banish forever the old dualism."[22]

But the books which Einstein himself credited as having been the most influential on him in his youth were Ernst Mach's *Theory of Heat* and *Science of Mechanics*. That author was motivated by the same Enlightenment animus and employed the same language. In the *Science of Mechanics*, Mach exclaims: "Science cannot settle for a ready-made world view. It must work toward a future one...that will not come to us as a gift. We must earn it! [At the end there beckons] the idea of a unified world view, the only one consistent with the economy of a healthy spirit."[23]

Indeed, in the early years of this century, German scientists were thrashing about in a veritable flood of publications that called for the unification or reformation of the "world picture" in the very title of their books or essays. Planck and Mach carried on a bitter battle, publishing essays directly in the *Physikalische Zeitschrift* with titles such as "The Unity of the Physical World Picture." Friedrich Adler, one of Einstein's close friends, wrote a book with the same title, attacking Planck. Max von Laue countered with an essay he called "The Physical World Picture." The applied scientist Aurel Stodola, Einstein's admired older colleague in Zurich, corresponded at length with Einstein on a book which finally appeared under the title *The World View of an Engineer*. Similarly titled works were published by other collaborators and friends of Einstein, such as Ludwig Hopf and Philipp Frank.

Perhaps the most revealing document of this sort was the manifesto published in 1912 in the *Physikalische Zeitschrift* on behalf of the new *Gesellschaft für positivistische Philosophie*, composed in 1911 at the height of the *Weltbild* battle between Mach and Planck. Its declared aim was nothing less than "to develop a comprehensive *Weltanschauung*," and thereby "to advance toward a noncontradictory, total conception [*Gesamtauffassung*]." The document was signed by, among others, Ernst Mach, Josef Petzoldt, David Hilbert, Felix Klein, George Helm, Albert Einstein (only just becoming more widely known at the time), and that embattled builder of another world view, Sigmund Freud.[24]

It was perhaps the first time that Einstein signed a manifesto of any sort. That it was not a casual act is clear from his subsequent, persistent return to the same theme. His most telling essay was delivered in late 1918, possibly triggered in part by the publication of Oswald Spengler's *Decline of the West*, that polemic against what Spengler called "the scientific world picture of the West." Einstein took the occasion of a presentation he made in honor of Planck (in *Motiv des Forschens*) to lay out in detail the method of constructing a valid world picture. He insisted that it was not only possible to form for oneself "a simplified world picture that permits an overview [*übersichtliches Bild der Welt*]," but that it was the scientist's "supreme task." Specifically, the world view of the theoretical physicist "deserves its proud name *Weltbild* because the general laws upon which the conceptual structure of theoretical physics is based can assert the claim that they are valid for any natural event whatsoever....The supreme task of the physicist is therefore to seek those most universal elementary laws from which, by pure deduction, the *Weltbild* may be achieved."[25]

There is of course no doubt that Einstein's work during those years constituted great progress toward this self-appointed task. In the developing relativistic *Weltbild*, a huge portion of the world of events and processes was being subsumed in a four-dimensional structure which Hermann Minkowski in *"Raum und Zeit"* (1908) named simply *"die Welt"*—the world conceived as a Parmenidean crystal made of world lines, in which changes, such as motions, are largely suspended. In this world, the main themata are those of constancy and invariance, determinism, necessity, and completeness.

Leaving Out Not a Single Event

Typically, it was Einstein himself who knew best and recorded frequently the limitations of his work. Even as special relativity began to make converts, he announced that the solution was quite incomplete because it applied only to inertial systems and left out entirely the great puzzle of gravitation. Later he worked on removing the obstinate dualities, explaining for example that "measuring rods and clocks would have to be represented as solutions of the basic equation...not, as it were, as theoretical self-sufficient entities." This he called a "sin" which "one must not legalize." The removal of the sin was part of the hoped-for perfection of the total program, the achievement of a unified field theory in which "the particles themselves would *everywhere* be describable as singularity-free solutions of the complete field-equations. Only then would the general theory of relativity be a *complete* theory."[26] Therefore the work of finding those most general elementary laws from which by pure deduction a single, consistent, and complete *Weltbild* can be won had to continue.

There has always been a notable polarity in Einstein's thought with respect to the completeness of the world picture he was seeking. On the one hand, he insisted from beginning to end that no single event, individually considered, must be allowed to escape from the final grand net. We have already noted that in the Oxford lecture of 1933 he is concerned with encompassing the "totality of experience," and he declared the supreme goal of theory to be "the adequate representation of any content of experience."[27] He even goes beyond that: toward the end of his lecture he reiterates his old opposition to the Bohr-Born-Heisenberg view of quantum physics and declares, "I still believe in the possibility of a model of reality, that is to say a theory, which shall represent the events themselves [*die Dinge selbst*] and not merely the probability of their occurrence. Writing three years later he insists even more bluntly:

> But now, I ask, does any physicist whosoever really believe that we shall
> never be able to attain insight into these significant changes of single sys-

tems, their structure, and the causal connections, despite the fact that these individual events have been brought into such close proximity of experience, thanks to the marvelous inventions of the Wilson-Chamber and the Geiger counter? To believe this is, to be sure, logically possible without contradiction; but it is in such lively opposition to my scientific instinct that I cannot forgo the search for a more complete mode of conception.[28]

Yet even while Einstein seemed anxious not to let a single event escape from the final *Weltbild*, he seems to have been strangely uninterested in nuclear phenomena, that lively branch of physics which began to command great attention precisely in the years Einstein started his own researches. He seems to have thought that these phenomena, in a relatively new and untried field, would not lead to the deeper truths. And one can well argue that he was right; not until the 1930s was there a reasonable theory of nuclear structure, and not until after the big accelerators were built were there adequate conceptions and equipment for the hard tests of the theories of nuclear forces.

Einstein's persistent pursuit of a fundamental theory, one so powerful that it would not leave out a single datum of experience and yet excluded nuclear phenomena, can be understood as a consequence of his suspension of disbelief of an extraordinary sort. It is ironic that, as it turned out, even while Einstein was trying to unify the two long-range forces (electromagnetism and gravitation), the nucleus was harboring two additional fundamental forces and, moreover, that after a period of neglect, the modern unification program, two decades after Einstein's death, began to succeed in joining one of the nuclear (relatively short-range) forces with one of the relatively long-range forces (electromagnetism). In this respect, the labyrinth through which the scientists have been moving appears now to be less symmetrical than Einstein had thought it to be.

For this and similar reasons, few of today's working researchers consciously identify their drive toward "grand unification" with Einstein's. Their attention is attracted by the thematic differences, expressed for example by their willingness to accept a fundamentally probabilistic world. And yet the historian can see the profound continuity. Today, as in Einstein's time and indeed that of his predecessors, the deepest aim of most fundamental research is still to achieve a logically unified and parsimoniously constructed system of thought that will provide the conceptual comprehension, as complete as humanly possible, of the scientifically accessible sense experiences in their full diversity. This ambition embodies a *télos* of scientific work itself, and it has done so since the rise of science in the Western world. Most scientists, working on small fragments of the total structure, are as unselfconscious about their participation in that grand monistic task as they are about, say, their monotheistic assumption, which remains central to their personal belief system without demanding explicit avowal. Indeed, Joseph Needham may well be right that the development in the West of the concept of a unified natural science depended on the preparation of the ground through monotheism, so that one can understand more easily the reason that modern science arose in seventeenth-century Europe rather than, say, in China.

Thematic Pluralism and the Direction of Progress

Diversity in the spectrum of themata held by individual scientists, and overlap among these sets of themata: this formula seems to me to answer the question why the preoccupation with the eventual achievement of a unified world picture did not lead physics to a totalitarian disaster, as an Ionian Fallacy by itself could well have done. At every step, each of the various world pictures in use is considered a preliminary version, a premonition of the Holy Grail. Moreover, each of these various hopeful but incomplete world pictures that guide scientists at a given time is not a seamless, unresolvable entity (unlike a "paradigm"). Nor is each completely shared even within a given subgroup. Each member of the group operates with a specific spectrum of separable themata, some of which are also present in portions of the spectrum in rival world pictures. Einstein and Bohr agreed far more than they differed, but they did have profound thematic differences. Moreover, most of the themata current at any one time are not new—they very rarely are—but are adopted from predecessor versions of the *Weltbild*, just as many of them would later be incorporated in subsequent versions of it. Einstein freely called his project a "Maxwellian program" in this sense.[29]

It is also for this reason that Einstein saw himself with characteristic clarity not at all as a revolutionary, as his friends and his enemies so readily did. He took every opportunity to stress his role as a member in an evolutionary chain. Even while he was working on relativity theory in 1905, he called it "a modification" of the theory of space and time. Later, in the face of being acclaimed as the revolutionary hero of the new science, he insisted, as in his King's College lecture: "We have here no revolutionary act but the natural development of a line that can be traced through centuries." Relativity theory, he held, "provided a sort of completion of the mighty intellectual edifice of Maxwell and Lorentz."[30] Indeed, he shared quite explicitly with Maxwell and Lorentz some fundamental presuppositions, such as the need to describe reality in terms of continua (fields), even though he differed from them completely with respect to others, such as the role of a plenum.

On this model of the role which the thematic component plays in the advancement of science, we can understand why scientists need not hold substantially the same set of beliefs either to communicate meaningfully with one another in agreement or disagreement or in order to contribute to cumulative improvement of the state of science. Their beliefs have considerable fine structure; and within that structure there is room both for thematic overlap and agreement, which generally have a stabilizing effect, and for intellectual freedom, which may be expressed as thematic disagreements. Innovations emerging from the balance, even the "far-reaching changes," as Einstein termed the contributions of Maxwell, Faraday, and Hertz, require neither from the individual scientist nor from the scientific community the kind of radical and sudden reorientation implied in such currently fashionable language as revolution, Gestalt switch, discontinuity, incommensurability, conversion, etc. On the contrary, the innovations are coherent with the model of evolutionary scientific progress to which Einstein himself explicitly adhered and which emerges also from the actual historical study of his scientific work.

Thus, major scientific advances can generally be understood in terms of an evolutionary process that involves battles over only a few but by no means all of the recurrent themata. The work of scientists acting individually or a group, seen synchronically or diachronically, is not constrained to the phenomenic-analytic plane alone; it is, rather, an enterprise whose saving pluralism resides in its many internal degrees of freedom. Therefore we can understand why scientific progress is often disorderly, but not catastrophic; why there are many errors and delusions, but not one great fallacy; and how mere human beings, confronting the seemingly endless, interlocking puzzles of the universe, can advance at all—as advance they have, if not all the way to the Elysium of Einstein's hope, the single world conception that grasps the totality of phenomena.

References

1. George Sarton, *The Study of History of Science* (1936; rpt. New York: Dover, 1957), p. 5.
2. Pierre Duhem, The Aim and Structure of Physical Theory (New York: Atheneum, 1962), pp. 176–278.
3. P.A. M. Dirac, *Principles of Quantum Mechanics* (Oxford: Clarendon Press, 1958), p. 312.
4. Victor Weisskopf, "Of Atoms, Mountains, and Stars: A Study in Qualitative Physics," *Sci-

ence, vol. 187, no. 4177 (1975): 605–612.

5. Robert G. Sachs, "Structure of Matter: A Five-Year Outlook," *Physics Today*, vol 32, no. 12 (December 1979): 27.

6. S.G. Brush, "Review," *American Journal of Physics*, 45 (1977): 687–688.

7. Usually translated as "On the Method of Theoretical Physics," this was Einstein's Herbert Spencer lecture, given at Oxford on June 10, 1933. The original manuscript of Einstein's essay was in German and has been published in his collection *Mein Weltbild* (Frankfurt am Main: Ullstein, 1977), pp. 113–119. An English translation appeared as a small booklet by Oxford University Press in 1933; but it left a good deal to be desired, and a different English translation was prepared (by Sonja Bergmann) when Einstein later published a collection of his essays in *Ideas and Opinions* (New York: Crown, 1954). The translation appears on pp. 270–276. In quoting from Einstein's essay and from his other writings, I refer to the pages of the English translations in *Ideas and Opinions*, but I have gone back to the corresponding original German essays and corrected the published English versions as necessary. In this connection, I wish to acknowledge with thanks the permission of the Estate of Albert Einstein to quote from Einstein's writings.

8. Philipp Frank, *Einstein: His Life and Times* (New York: Knopf, 1947), p. 217.

9. Quoted in Richard K. Gehrenbeck, *C.J. Davisson, L.H. Germer, and the Discovery of Electron Diffraction*, Ph.D. dissertation, University of Minnesota, 1973, pp. 343–344.

10. After the observation of the bending of light, published in November 1919, Einstein amended this sentence for the edition printed in 1920: now there remained only one consequence drawn from the theory which had not been observed (the red shift of spectral lines); but, he added, "I do not doubt at all that this consequence of the theory will also find its confirmation soon." Albert Einstein, *Über die spezielle und die allgemeine Relativätstheorie*, 7th ed. (Braunschweig: Vieweg, 1920), p. 70.

11 E.g., Einstein, *Ideas and Opinions*, p. 272. The quotations in the next six paragraphs are from the same source, pp. 273–276.

12. In the same essay, "Physik und Realität," 1936, translated as "Physics and Reality," in Einstein, *Ideas and Opinions*, p. 294.

13. In Pierre Speziali, ed., *Albert Einstein, Michele Besso, Correspondence 1903-1955 (Paris: Hermann, 1972), p. 527.*

14. Isaiah Berlin, *Concepts and Categories* (New York: Viking Press, 1979), p. 159.

15. Albert Einstein, in *Albert Einstein: Philosopher-Scientist*, ed. Paul A. Schilpp (Evanston, Ill.: Library of Living Philosophers, 1949), p. 53.

16. Albert Einstein, "Induktion und Deduktion in der Physik," *Berliner Tageblatt* (Supplement), December 25, 1919.

17. Einstein, in *Albert Einstein*, pp. 673–674. See also p. 678: "categories are necessary as indispensable elements of thinking."

18. A brief survey of thematic analysis and some case studies are provided in Gerald Holton, *Thematic Origins of Scientific Thought: Kepler to Einstein* (Cambridge, MA: Harvard University Press, 1988), and in Gerald Holton, *The Scientific Imagination: Case Studies* (New York: Cambridge University Press, 1978).

19. Max Planck, "Verhältnis der Theorien zueinander," in *Die Kultur der Gegenwart*, ed. Paul Hinneberg, Part III, vol. 1 (Leipzig: B.G. Teubner, 1915), p. 737.

20. A. Einstein, *Aether und Relativitätstheorie* (Berlin: Julius Springer, 1920), p. 14.

21. J. T. Merz, *A History of European Thought in the Nineteenth Century* (London: William Blackwood & Sons, 1904), I, 251–252.

22. Ludwig Büchner, *Kraft und Stoff: Empirisch-naturphilosophische Studien*, 9th ed. (Leipzig: Theodor Thomas, 1867), p. 89.

23. Ernst Mach, *Die Mechanik in ihrer Entwicklung, historisch-kritisch dargestellt*, 2nd ed. (Leipzig: F. A. Brockhaus, 1889), pp. 437–438.

24. Cf. "Aufruf," *Physikalische Zeitschrift*, 13 (1912): 735–736; and Friedrich Herneck, "Albert Einstein und der philosophische Materialismus," *Forschungen und Fortschritte*, 32 (1958): 206. For an analysis of the document, and its sources and influence, see chapter 1,

"Ernst Mach and the Fortunes of Positivism" in G. Holton, *Science and Anti-Science* (Cambridge, MA: Harvard University Press, 1993).

25. Albert Einstein, "Motiv des Forschens;" a somewhat loose English translation was published in Einstein, *Ideas and Opinions*, pp. 224–227.

26. In *Albert Einstein*, pp. 59–61, 81; emphases in original, In the 1933 Oxford lecture, Einstein raises this whole problem only gently, and at the end, by saying: "Meanwhile the great stumbling block for a field theory of this kind lies in the conception of the atomic structure of matter and energy. For the theory is fundamentally nonatomic insofar as it operates exclusively with continuous functions of space" (p. 275), unlike classical mechanics, which by introducing as its most important element the material point does justice to an atomic structure of matter. He does see a way out:

> For instance to account for the atomic character of electricity field equations need only lead to the following conclusion: The region of three-dimensional space at whose boundary electrical density vanishes everywhere always contains a total electrical charge whose size is represented by a whole number. In the continuum theory, atomic characteristics would be satisfactorily expressed by integral laws without localization of the entities which constitute the atomic structure. (p. 276).

In referring to total electric charge whose size is represented by a whole number, he points of course to the result of R. A. Millikan's work.

27. Einstein, *Ideas and Opinions*, p. 272. The phrase was translated in the first English version of the 1933 lecture delivered by Einstein as "the adequate representation of a single datum of experience."

28. Einstein, *Physik und Realität*, p. 318.

29. The case is quite general. Thus, Kepler's world was constructed of three overlapping thematic structures, two ancient and one new: the universe as theological order, the universe as mathematical harmony, and the universe as physical machine. Newton's scientific world picture clearly retained animistic and theological elements. Lorentz's predominantly electromagnetic world view was really a mixture of Newtonian mechanics as applied to point masses, determining the motion of electrons, and Maxwell's continuous-field physics. Ernest Rutherford, writing to his new protégé, Niels Bohr, on March 20, 1913, gently scolds him: "Your ideas as to the mode of origin of spectra in hydrogen are very ingenious and seem to work well; but the mixture of Planck's ideas [quantization] with the old mechanics makes it very difficult to form a physical idea of what is the basis of it." In fact, of course, Bohr's progress toward the new quantum mechanics via the correspondence principle was a conscious attempt to find his way stepwise from the classical basis.

30. Einstein, "On the Theory of Relativity," in *Ideas and Opinions*, p. 246.

Of Physics, Love,
and Other Passions:
The Letters of Albert and Mileva

T he popular image of the scientist as *dispassionate*, and therefore somehow less than human, is undoubtedly bolstered by the convention according to which modern scientists remove their personal hopes and desires as much as possible from their published research results. Editors of professional journals apparently frown on the use of the word "I" in articles. There are in fact good reasons for the apparent impersonality of contributions to the scientific literature; one aim of published science is to minimize any obstacle against building consensus, and thus hasten the conversion of private science into public science.

But when historians of science look at the stages of a scientist's work prior to publication, a different picture emerges. Depending on the case, dispassion may be the least of the virtues in evidence. The very motivations for persisting in the exhausting pursuit of a difficult problem would be inexplicable without the operation of a whole range of emotions that banal opinion would grant to everyone except scientists. And when it comes to studying the private letters of certain scientists (such as Schrödinger), the evidence of passions ruling both the personal and scientific life can reach volcanic proportions.

One must approach such a case with caution because of the difficulty of trying to put oneself into the frame of mind in which very personal letters were written at some point in the distant past. This problem was put very well by A. S. Byatt in her novel *Possession—A Romance* (1990), a fascinating concoction centering on the discovery of love letters between the fictional Randolph Henry Ash, mid-nineteenth-century poet and ama-

*Mileva Marić in
about 1896.*

teur scientist, and the mysterious writer, Christabel La Motte. The book
proposes that two late-twentieth-century English literature scholars, Ro-
land Mitchell and Maud Bailey, discover that correspondence, with its record of the romance, the mysterious ending of it, the question of whether
there was a child and what became of it. Byatt indicates the problem for
the contemporary scholar. Maud says: "I have been trying to imagine him.
Them. They must have been—in an extreme state....We are very knowing.... We know we are driven by desire, but we can't see it as they did,
can we? We never say the word Love, do we—we know it's a suspect ideological construct—especially Romantic Love—so we have to make a real effort of imagination to know what it felt like to be them, here, believing in
these things—love—themselves—that what they did mattered—"

Let's make that effort, for a couple of young people who longed for
each other as well as for a life in science.

Mileva's family in about 1883. Right to left: Mileva; her father, Miloš; his adopted daughter Nana and wife, Marija, with baby daughter Zorka. (From Desanka Trbuhović-Gjurić, Im Schatten Albert Einsteins, Paul Haupt Verlag, Bern, 1988; reproduced with permission of the publisher.)

Mileva

Mileva Marić was born in 1875. Her father, Milos, was a Serb, and her mother, Marija Rûzić, came from a Montenegran family. They were living in Vojvodina, then part of southern Hungary, a few dozen miles northwest of Belgrade, an area that had initially been settled by farmer-warriors to protect Christian Europe from the Turks. Mileva's was a rather well-to-do family, her father being an official in the state bureaucracy. From earliest

*Einstein in about
1895–96. (Cour-
tesy of Elena
Sanesi.)*

times, she showed that she would be a serious and dedicated young woman. For example, during her high school years, she sought and received special permission to attend an all-male *Gymnasium* to take two years of physics—a course which otherwise was not available to girls generally. It is reported that she got top grades in physics as well as in mathematics.

It was very difficult for a woman to go on to get higher education in most parts of Europe. The first European country to admit women to its universities was France in 1863, followed by the University of Zurich in Switzerland in 1865, with other Swiss universities following a few years later. So if you were French or, like Maria Sklodowska (later Marie Curie), comfortable with the French language, you would naturally move to Paris for university training. If, like Mileva, you were fluent in German, it was more natural to go to Switzerland. Indeed, as John Stachel points out, "most of the women entering Swiss higher schools during the

Einstein's parents, Hermann and Pauline (née Koch).

nineteenth century were non-Swiss, a large proportion of them being of Slavic background."[1]

In 1894, Mileva came to Switzerland, first taking the last two years of secondary school, then doing a summer semester in medical studies at the University of Zurich in 1896, and finally, in the fall of that year, enrolling in the Swiss Federal Polytechnic School (called "Poly" for short, or "ETH," for Eidgenössische Technische Hochschule).

From the testimony of her fellow students and those who knew her soon after, she must have been a formidable young woman, with a fierce dedication to the study of physics. In conversation, she tended to be quiet. In appearance she was darkly handsome. Having been born with a congenital hip dislocation from which her sister also suffered, she walked with a slight limp. One is reminded of studies such as that of sociologist Anne Roe, who found that scientists, far beyond the usual proportion in the general population, had periods of illness or physical handicap in childhood. Some psychologists[2] have related this fact to the prominent incidence of introversion among young persons hoping to become scientists—all such characteristics helping the youth to resist early peer pres-

Mileva and Albert Einstein with their son Hans Albert in 1904.

sure, which generally was away from science as a career. A scientist of whom we shall hear more presently described another mechanism serving the same end when he wrote in his autobiography that he had passed through an early period of "deep religiosity...which...found an abrupt ending at the age of twelve, through the reading of popular scientific books." And he added, "It is quite clear to me that the religious paradise of youth, which was thus lost, was a first attempt to free myself from the chains of the 'merely personal,' from an existence which is dominated by wishes, hopes, and primitive feelings."

A Love Letter

The letter from Mileva to Albert whose first page is shown at left was written in 1901, probably soon after the Como expedition. The translation cannot do justice to the endearing diminutives (Schatzerl, Brieferl, Weiberl, Manderl, even Gotterl), which are reminiscent of the lovers Papageno and Papagena in Mozart's Magic Flute. The following is a partial translation:

Dearest Little Treasure!

Now I've also received your second little letter and I'm so happy, beyond all measure. How dear you are, oh how I'll kiss [busseln] you, I can hardly wait for the end of the week, when you come.... If you come Saturday you might be able to sleep at our (rooming house), because someone is leaving Friday. I'll ask [the landlady], she'll do it for me if it's possible. Till then I'll work very hard so I'll be quite free to enjoy your company—God [Gotterl], how beautiful the world will look when I'm your little wife, you'll see there'll be none happier in the whole world, and then it must be the same for the man [Manderl].

Stay well my sweet little treasure and come in good cheer at week's end to
Your woman [Weiberl].

Wedding portrait of Mileva and Albert in Bern, January 1903.

What were Mileva's career interests? One must remember that the chance of obtaining eventually a university professorship in any science in the 1890s and early 1900s was very small even if one completed the regular academic university preparation. Such positions were few (54 in physics at all university ranks in Austria-Hungary in 1909); openings occurred rarely, and the road to them was long. Whether for this reason or because her ambitions lay elsewhere, Mileva enrolled in the Poly as a student in Section VIA, which trained teachers of physics and mathematics primarily for secondary schools. Once more, she was the only female student in her class—to this day a difficult thing to be.

As it happened, among the handful of other students enrolling at the same time in the same section and for the same purpose, there was a bright young seventeen-year-old chap, about 3½ years her junior. He had also come to Switzerland from abroad, and also had already shown that he was devoted to the study of science. His name was, of course, Albert Einstein.[3] Through their letters we can begin to glimpse the story of their love, their physics, and their other passions.

A sketch. In February 1902 Albert sent Mileva this sketch he had made of his furnished room in Bern. His explanation (below) of the symbols is full of diminutives and dialect expressions, as is so much of their correspondence. The last item (Γ) is perhaps a picture of Mileva.

B = *Betterl [little bed]*
b = *Bilderl [little picture]*
d = *Deckerl [little mat]*
gS = *grossartiger Sessel [terrific easy chair]*
gΣ = *grossartiger Spiegel [terrific mirror]*
J = *Johonzel [Little Johnny (himself)]*
K = *Kasten [chest]*
κ = *Kanapyen [sofa]*
κΣ = *kleiner Spiegel [small mirror]*
N = *Nacht-Topf & Tisch [chamberpot & table]*
F = *Fensterl [little window]*
O = *Ofen [stove]*
S = *Stühlerl [little stool]*
T = *Thüre [door]*
τ = *Tisch [table]*
υ = *nichts [nothing]*
U = *Uhrerl [little clock]*
Γ = *Gell da schaugst! [I'll bet you're surprised!]*

Mileva Einstein and her sons, Eduard (left) and Hans Albert, around 1914.

Young Einstein came to the Poly after a final year of secondary-school education at the Kantonsschule in Aarau, Switzerland. He had been there as a boarder in the home of the teachers of Greek and of History at the school, Jost Winteler and his wife Pauline. They had seven children and kept a warm and comfortable house in which Einstein flourished. Einstein was soon as devoted to the Winteler parents as to his own, or perhaps even more. (Eventually, one of the sons married Einstein's sister, Maja, and one of the daughters married Einstein's closest friend, Michele Besso.) Moreover, while at Aarau, Albert and the youngest of the daughters, Marie, were greatly attracted to each other, and both sets of parents thought they would eventually make a good couple. In fact, if we look into the first volume of the published correspondence of Einstein (*The Collected Papers of Albert Einstein,* Princeton University Press, 1987), one of the earliest letters (November 1896) is from Marie Winteler to him, starting with the appellation "Beloved Treasure" ("Geliebter Schatz") and ending with "In deep love, your Mariechen." It's a rather desperate love letter from one adolescent to another, in which she pleads with him not to break off the correspondence.

Hans Albert and Eduard (right) with Einstein in the mid-1920s.

The Einstein Archive

Let me interpolate here to say something about the Einstein Archive that is the basis of much that follows. By the will of Albert Einstein, all correspondence, books, manuscripts, etc., in his possession at his death were eventually transferred to the Hebrew University Library in Jerusalem, and most of the correspondence is being published, several volumes out of an initially projected total of thirty having already appeared. The archive is a real treasure trove, since Einstein corresponded not only with most of the major and many of the minor scientists of his time, but also with people in political life, from Gandhi and Roosevelt down to Stalin; with writers and musicians, sages and cranks, and the wretched of this earth. In a way, the archive is a window on a good deal of the triumphs and terrors of the twentieth century, and it is now being used widely by scholars in the most diverse fields.

The fact that it exists in a form that can be readily used by scholars is in some part due to my stumbling on the documents almost by accident. After Einstein's death in 1955, I was asked to participate in a memorial

Einstein and his son Hans Albert in 1904 or early 1905.

symposium. To my surprise, I found that very little had been written by historians of physics on the impact of Einstein's work, and I thought that my contribution might be to begin such an assessment, based on whatever documents might be available. Philipp Frank, then a colleague in my department and the good friend and biographer of Einstein, recommended that I examine whatever manuscripts or letters I might find still at the Princeton Institute for Advanced Study or at Einstein's old house on Mercer Street.

Arriving in Princeton, I was told that all available materials might be in some files in the big basement safe room at the Institute's main building, Fuld Hall. I was not prepared for what I found there. In the darkness, against the background of more than twenty tall metal file cabinets, was Einstein's secretary since 1928, Helen Dukas. The goose-neck table lamp on her small desk provided the only illumination. She was busy, responding to letters still addressed to Einstein or requesting reprint rights. The whole scene reminded me of Juliet in the crypt. When I asked whether there were documents illustrating Einstein's way of thinking and working in physics, and his interaction with his colleagues, she opened some of the file cabinets. Arranged in a way accessible to her but perhaps to no one else, here were the many thousands of letters, drafts of answers, published and unpublished manuscripts, etc., that had been carefully preserved by his devoted helpers over the years. After Einstein's immigration to the United States in 1933, most of the material then in existence had been spirited out of Nazi Germany, with the aid of the French Embassy in Berlin, and sent to the U.S. in a so-called diplomatic pouch arrangement.

Looking at this chaotic collection, and getting from Helen Dukas the distinct impression that more was stored at Einstein's house in Princeton in which she and his stepdaughter Margot were still living, I felt compelled to get this mountainous material organized into a workable archive. In those days, one of the large private foundations acted quickly and sensibly on requests for aid in cultural matters of this sort. So I was able to hire Helen and some physics graduate students from Princeton, and arranged for periodic visits, to help convert the files into an archive, and to prepare a catalogue raisonné summarizing each of the documents. Eventually the archive turned out to contain some 45,000 documents, including those added since 1955 by gift or purchase. Among those are the copies of the fifty-four letters exchanged between Einstein and Marić between 1897 and 1903 (forty-three of his, eleven of hers), now published in the *Collected Papers*, which were made available in 1986 by the family of Einstein's first son, Hans Albert.

The Correspondence

Now back to the letters of the late 1890s. The first surviving one between Mileva and Albert is hers of 20 October 1897 (about a year after Marie's). At the Poly, she and Albert had been taking mostly the same required courses—although he also took classes in business, banking and in insurance-statistics, to be on the safe side in case he should be needed to help run the family business (electrical engineering). But now, during her second year, Mileva had taken a leave for a few months to attend lectures in physics and mathematics at the University of Heidelberg. In that first mailing we have from her hand, she addresses the young man by the formal *Sie,* rather than the informal *Du.* Perhaps because of the ambiguity of just how to address each other, her letter has neither initial nor final salutation. It is rather cool, evidently in reply to a four-page letter she had received from him a while earlier, one of the many which have not been preserved. She writes that on a visit back home she had told her Papa all about Albert: "You absolutely must come [home] with me some day. You would enjoy yourself splendidly." She adds, as if it were a non sequitur, that one of their mutual acquaintances seems to have become involved in a romance, and she concludes: "What's the point of him falling in love nowadays, anyway? It's such an old story. . . ." She finishes with a brief remark about a lecture she has just heard on the kinetic theory of gases, based on the then-still controversial concept that gases consist of atoms.

Chronologically, the next document in the archives is an envelope Albert had addressed to Mileva. But the envelope is empty—a reminder that as scholars or snoops we are restricted to whatever scraps may be left over. Six weeks later, there is a letter from Albert to Mileva. He also addresses her quite formally: "Geehrtes Fräulein." He looks forward to her impending return to Zurich. Because she has fallen a year behind her classmates there, he offers to make available his class notes so that she will be able to catch up. (When she later used them, she penned a correction of one of Einstein's small mistakes—the vessel on the right, not the left—and this correction, properly credited to her, has now appeared in Einstein's *Collected Papers.* The irony of it will become clear later.)

Now the friendship is turning into a romance amidst the experimental equipment in the physics lab of the Poly. They spend much of their time together studying for their courses, and doing the required experimental thesis project needed to obtain the diploma. During vacations, Albert often goes back to visit his family in Milan, where his father has tried to start a new business after an earlier failure of his company. On those occasions we have letters between them. For example, Albert writes in March 1899 that he realizes "how closely knit our psychological and physiological life is." He had shown a photograph of Mileva to his mother, and writes: "Your photograph had a great effect on my old lady"; she had studied it carefully, and "sends her greetings."

But Mother's greetings were not from the heart. Slowly Albert realizes his parents' hostility to this friendship, and at the same time, in reaction, he begins to see them in a new light. Early August 1899: "My mother and sister seem a little petty and philistine to me, despite the sympathy I feel for them. It is curious how gradually our lives change us in the very subtle shadings of our soul, so that even the closest family ties dwindle into habitual friendship, and deep inside one becomes mutually so incomprehensible that one is in no way capable of empathizing with the emotions that move the other." And, again, in the second letter of that month: "I manage to escape their mindless prattle....If only you were again with me for a while! We understand so well each other's dark souls, and also drinking coffee & eating sausages, etc......" (The six final periods are underlined.)

Electrodynamics of Moving Bodies

Even as the romance is getting more stable and serious, another thing comes to the foreground as well. In virtually each of his longer letters, aside from personal matters often written in his witty and colloquial German to which translations can hardly do justice, Einstein goes into details of the physics that is preoccupying him—not merely rapturous accounts of what he is reading in his textbooks or in classics of science, but new ideas. For example, in that second letter of August 1899: "[I] am now re-reading Hertz's "Propagation of Electric Force" with greatest care. The reason is that I did not understand Helmholtz's treatise....I am more and more reaching the conviction that the electrodynamics of moving bodies, as it is presented today, doesn't correspond to reality, but can be presented more simply. The introduction of the term 'ether' into the theories of electricity has led to the conception of a medium of whose motion one can speak but without, I believe, being able to connect a physical meaning with this way of speaking. I believe that electric forces are definable only for empty space, as Hertz also stressed." And so forth. This is six years before the third of his major papers of 1905, the one on relativity theory, "On the Electrodynamics of Moving Bodies," containing in its title the very phrase he uses here, and which had also been used by Hertz. But it is perhaps revealing of their relative interests that while Einstein devotes much of his letters to his new ideas in physics, in her reply she is totally silent on that subject, and comments only on family matters and the like. On the whole, that asymmetry is true for most of their preserved exchanges. In the relatively few letters of Mileva during the period under discussion, she never responds to his arguments about physics, nor does she otherwise write of scientific problems except to refer occasionally to a lecture she has heard or to a science book.

But Einstein continues to present his scientific ideas to her. Thus in his next letter (10 September 1899): "In Aarau [where he went to visit the

Wintelers, probably with his sister who was entering the girls' school there] I had a good idea for investigating what effect a body's relative motion with respect to the luminiferous ether has on the velocity of propagation of light in transparent bodies." It is apparently a reference to some variant of the Fizeau experiment of 1851. As Einstein wrote much later, the clues provided by the Fizeau experiment and stellar aberration, together with the old experiment of Faraday on electromagnetic induction, "were enough" to put him on the trail of relativity theory.

Other tantalizing hints of his reading and thinking are sprinkled throughout his correspondence. For example, in a letter two weeks after the one just quoted, he says to Mileva: "I also wrote to Professor Wien in Aachen about my work on the relative motion of the luminiferous ether against ponderable matter, the work which the 'boss' handled in such a negligent fashion." This presumably refers to his physics teacher, H. F. Weber. As one of Einstein's biographers, his son-in-law (Rudolf Kaiser, who published the book *Albert Einstein* under the pseudonym Anton Reiser in 1930, p. 52) put it, during his student days, Einstein "wanted to construct an apparatus which would accurately measure the earth's movement against the ether," but was not able to do so because "the skepticism of his teachers was too great, the spirit of enterprise too small."

Einstein goes on to say: "I read a very interesting paper by Wien from 1898 on this subject." That paper, on theories of the motion of the presumed ether, described thirteen of the most important experiments trying to detect the earth's motion through that ether. The vast majority of them, ten, had negative results, and among those, indeed the last of them, was the Michelson-Morley experiment which later was elevated by some commentators as the only experiment of significance in the genesis of relativity theory—a position Einstein repeatedly disavowed.

In the same letter (28 September 1899): "I tend to brood too much—in short I see and feel the absence of your beneficent thumb under which the boundaries are kept." It is a puzzling and possibly revealing sentence, followed by other hints of his complex state of mind: "Don't worry about my now going so often to Aarau. The critical daughter with whom I was so madly in love four years ago [he is referring to Marie Winteler] is coming home. For the most part I feel quite secure in my high fortress of calm. But if I saw her again a few times, I would certainly go mad, that I know & fear like fire….A thousand hearty greetings from your Albert."

As we turn to the letters of 1900, something new appears. They have agreed on an important step: to address each other finally by the familiar *Du* (thou), and to refer to each other by nicknames, part of the usual scenario of developing affections: "Mein liebstes Doxerl [doll]," but also

"Meine süsse Kloane [little one]," "Meine liebe Miez [pussycat]," and variants on all of these. He in turn calls himself, and is often addressed by her as, Johannesl or Johanzel. In a letter to a friend, Mileva calls Einstein her "Hauptperson," her "significant other," as we now would put it.

By the end of July, a major turning point event occurs for each of them, and therefore for their relationship. A letter from the university authorities of 27 July 1900 brings the news that Albert Einstein has passed his diploma examination. But Mileva, taking the examination at the same time, failed it. (Her grades were quite good in theoretical and practical physics, but not good enough in mathematics.) From other sources we know that, tired and depressed, Mileva decided to leave, to visit her parents. A day or two later, Einstein wrote her from a summer vacation place in Switzerland where once again he had joined his parents and sister. It is a dramatic, even theatrical letter, which shows how his emotional life is rearranging itself. Let me quote it at length:

> I arrived in Sarnen the day before yesterday as planned.... We were met by Mama, Maja, and a carriage.... Maja took this opportunity to say [privately] that she had not dared to report anything about the "Dockerl" affair, and she asked me to "go easy" on Mama.... So we arrive home [at the hotel] and I go into Mama's room, only the two of us. First I must tell her about the exam [that he had passed, and Mileva failed], then she asks me quite innocently: "So, what will become of Dockerl now?" "My wife," I said just as innocently, but prepared for the expected "scene."

> That followed immediately. Mama threw herself onto her bed, buried her head in the pillows, and wept like a child. After regaining her composure, she immediately shifted to a desperate attack: "You are ruining your future and barring your career.... No decent family will have her. ...If she gets pregnant, you'll be in a mess." With this last outburst, I finally lost my patience. I vehemently denied the suspicion that we had been living in sin, and scolded her mightily....

> The next day things were better, largely because, as she said herself, "If they have not yet been intimate (which she had greatly feared) and are willing to wait longer, then ways and means can be found." The only thing that is out of the question for her is that we want to remain together always. The attempts at changing my mind came in expressions such as "Like you, she is a book—but you ought to have a wife." "By the time you are 30, she is an old witch," etc.... If only I could be with you again soon in Zurich, my little treasure! A thousand greetings and the biggest kisses from your Johannesl.

Stubborn as a Mule

He has now assumed the role as Mileva's defender and comforter (though one might doubt whether a letter like that would really soothe her). A few days later, he adds: "Mama and Papa are quite phlegmatic types and have less stubbornness in their entire bodies than I have in my little finger." On that he was probably quite right. When asked later why it was he who came to find the relativity theory, he gave two answers: that he had curiosity and the stubbornness of a mule, God's only gift to him; and that he continued to ask questions that children eventually are taught not to ask.

The more Albert now experienced his family's rejection of Mileva, the more he was drawn toward her and away from them. He writes on 1 August 1900, "I long terribly for a letter from my beloved witch. I can hardly comprehend that we are for such a long time separated—only now do I see how terribly dear you are to me! Indulge yourself completely so you will become a radiant little darling and as mad as a street urchin [*Gassenbub*]." And two weeks later (14 August 1900), "How was I able to live alone before, my little everything? Without you I lack self-confidence, passion for work, and enjoyment of life—in short, without you my life is no life."

But as one would expect, the escape from the merely personal is always at hand for him. On 30 August 1900: "My only diversion is studying, which I am pursuing with redoubled effort, and my only hope is you, my dear, faithful soul." And 13 September 1900: "Boltzmann is quite magnificent.... He is a masterful writer. I am firmly convinced of the correctness of the principle of the theory, i.e., I am convinced that in the case of gases we are really dealing with discrete mass points of definite, finite magnitude that move according to certain conditions." This is of course a basis for his 1905 paper with the proof of the molecular hypothesis through Brownian motion.

On 19 September 1900, Einstein shows once more he is anxious to escape a "philistine" existence—in his wide reading he was drawn to Schopenhauer, who for young people at the time was a strong voice against following the crowd, and in later letters Einstein often referred to himself as a gypsy at heart. Now he paints an optimistic picture for the two young people in love (19 September 1900): "No matter what may happen we'll have the most delightful life in the world. Pleasant work and being together—and what's more, we both are now our own masters and stand on our own two feet and can enjoy our youth to the utmost. Who could have it any better? When we have saved up enough money, we shall buy ourselves bicycles and take a bike tour every couple of weeks."

"Our Work on Relative Motion"

By late October 1900, Einstein must have rejoined her in Zurich, and there are no more letters between them until spring 1901, when he visits his parents again. She has been left alone, studying to take that examination for the second time. It is a low point for her; perhaps it is dawning on her that she will fail again and will never get her degree. And at that moment he writes a letter (27 March 1901) from that hostile camp which contains a phrase on which some commentators fastened for a time. "My dear kitten. Many thanks for your letters and for all true love contained therein. I kiss you and hug you with all my heart for it.... You are and will remain a shrine for me to which no one has access; I also know that of all people you love me the most and understand me the best. I assure you that no one here would dare or even want to say anything bad about you. I'll be so happy and proud when we together shall have brought our work on relative motion victoriously to a conclusion!"

"*Our* work on relative motion." In their surviving correspondence, it is his only use of the word *our* in the context of the early labors that led to the relativity theory four years later; but it will occur again in other contexts, and requires some comments later to understand what may have been meant.

We come now to another key event in the life of these struggling young people. Now that Einstein had his diploma, he was hoping to find a job. His applications for assistantships to various physics professors in Switzerland, Germany, and Italy yielded no encouraging replies, but in mid-April 1901 he had good news. He was invited to be a substitute teacher for two months at the Technical High School in Winterthur, and there also was a letter from his friend Marcel Grossmann, saying that (thanks to the intervention of Marcel's father) Einstein might perhaps have a chance to get a position at the Swiss Patent Office in Bern. Strangely, his report to Mileva on all this is given only a few lines at the beginning of his letter to her. The main topic is a lengthy discussion of what he calls his new, "wonderful idea" to apply a theory of molecular forces to gases, followed by a long exposition of the physics involved.

A few days later, still in high excitement, he again tries to lift her darkening mood—she had even begun to look for a secondary school position in Zagreb, and incidentally she was also jealous of the attention he seems to have paid to some of her women friends. So on 30 April 1901, Einstein writes her: "My dearest little child [Kinderl]! I just don't let up. You absolutely must come to me in Como, you sweet little witch.... And I love you so much again! It was only out of nervousness that I was so mean to you.

You will hardly recognize me now that I have become so bright and cheerful, and I'm longing so much to see again my dearest Doxerl.... Come to me in Como and bring my blue morningrobe in which we can wrap ourselves up, and don't forget your binoculars." To be sure, this is immediately followed by his opinions on Boltzmann's gas theory. He believes that there is now on this subject "enough empirical material for our investigation," to be found in a book by O. E. Meyer, and asks her to check on it the next time she is in the library.

There is something a little frantic in this letter. Obviously, life is now looking quite different from their respective positions, and he is trying to cheer her up and to bind her to him. After some hesitation, Mileva accepts his invitation. She says she has just "received a letter from home today that has made me lose all desire, not only for having fun, but for life itself."

That excursion, later described in one of her letters to her friend Helene Savić, turned out to be perhaps the most important threshold point for the couple. They set forth north from spring-bound Como to the Splügen Pass, some 6,500 feet high, where the snow was in places still nearly twenty feet deep. They rented a sled and then went further on foot while more snow was falling. Mileva wrote later to a friend that she held on to him firmly throughout. It was a great success, "How happy I was again to have my darling for myself a little, especially because I saw that he was equally happy." It was all the more glorious for them to be together because for the next two months Einstein would be in Winterthur on his temporary teaching job.

"How Is the Boy?"

But perhaps the trip had gone all too well. A few weeks later (end of May 1901), one of Einstein's surviving letters to Mileva begins with his typical account of his almost physical euphoria on encountering in his reading a new physics problem—in this case the photoelectric effect. But then he almost seamlessly continues to write about the main new fact that has entered their lives: "I just read a wonderful paper by Lenard on the generation of cathode rays by ultraviolet light. [We know that following up on that experiment was responsible for the earliest of Einstein's major 1905 papers, the one usually referred to as his photon theory of light, for which he was given the Nobel prize in 1922.] Under the influence of this beautiful piece, I am filled with such happiness and joy that I absolutely must share some of it with you. Be of good cheer, dear, and don't fret. After all, I am not leaving you, and will bring everything to a happy conclusion. One just has to be patient! You'll see that one does not rest badly in my arms, even if things are beginning a little stupidly. How are you, darling? *How is the boy?*"

Mileva had discovered she was pregnant. This was surely not the first time that they had talked about it—he had visited her in Zurich sometimes during his work in Winterthur. But little did he seem to realize how this pregnancy would change their relationship. His first scientific paper, on capillarity, had been sent off the previous December. He was now launched as a publishing scientist (even though he thought later that his first published papers were quite insignificant). But for Mileva, becoming a science school teacher (not to speak of obtaining a university position of which she may only have dreamed) had already become very questionable. She was still studying for her second try at the examination in the teacher-track section of the Poly, which she was to take bravely at the end of July 1901, and, three months pregnant, would fail again with the same low total mark as before. And bearing an illegitimate child was of course in those days considered to be deeply shameful.

But at that point, in late spring 1901, Albert, despite his unquenchable optimism, was really not much more promising either—a twenty-two-year-old still without firm employment. His eventual job, a provisional appointment as Technical Expert Third Class at the Patent Office, actually did not materialize until about a year later. With little likelihood for even a menial permanent post, and with no moral support for his liaison from either his family or hers (there are even references to angry letters from Einstein's parents to Mileva's, which caused the predictable sorts of rows), he now faced, in straitlaced Switzerland, the prospect of becoming the impoverished father of an illegitimate child.

Here was a test of his fundamental morality. What would many a normal young man have done at that time under these conditions? For Einstein the answer was clear. A letter of 7 July 1901 to Mileva goes like this: "Rejoice now in the irrevocable decision I have made! About our future I have decided the following: I'll look for a position *immediately,* no matter how modest. [In fact, he seems to be thinking about going into an insurance firm as a clerk, or making do by giving hourly lessons.] My scientific goals and personal vanity will not prevent me from accepting the most subordinate role. As soon as I have such a position I will marry you and take you to live with me, without writing a word of it to anyone until everything has been settled. Then no one can cast a stone upon your dear head, and woe unto him who dares to set himself against you. When your parents and mine are presented with a *fait accompli,* they'll just have to reconcile themselves to it as best they can."

For whatever reason, this is not how it turned out. By November 1901, they write each other again because Mileva has gone home to her parents where she plans to have their baby. Albert is in yet another lowly temporary teaching position, in Schaffhausen, from which he soon manages to get himself dismissed for impudent behavior (such as asking for a better salary). Mileva writes 13 November 1901 to suggest that they keep the news of the pregnancy quiet: "I believe we should say nothing yet about Lieserl [to other friends]." The unborn child has at least been given a name; and it turned out later that the mother had guessed the gender correctly, although Albert replies, 12 December 1901, that: "Secretly [I] prefer to imagine a Hanserl." He hopes the Bern job will come through, but even if it does not, then "together, we'd surely be the happiest people on earth. We'll be students as long as we live and won't give a damn about the world.... The only problem that still needs to be solved is the question of how we can take our Lieserl to us; I do not want us to have to give her up. Ask your Papa, he's an experienced man...." And he adds, "She shouldn't be stuffed with cow's milk because it might make her stupid. Yours would be more nourishing, right?" And next, almost predictably: "I have had another rather obvious but important scientific idea about molecular forces...."

"So We Don't Become Old Philistines"

Five days later, 17 December 1901: "I am busily at work on an electrodynamics of moving bodies, which promises to be a capital piece of work. I wrote to you that I doubted the correctness of the idea about relative motion. But my reservations were based on a simple calculation error. Now I believe in it more than ever." And further on, "Give your mother my best greetings and tell her I am looking forward to the spanking with which she will do me honor some day." Two days later, 19 December, "Sorry, I forgot your birthday once again," and more thoughts on his electrodynamics. Nine days later, December 28, "When you become my dear little wife, we'll diligently work on science together so we don't become old philistines, right?"

Lieserl was born in January 1902, perhaps at Mileva's home in Novi Sad. Soon after he heard the news, Albert writes her (4 February 1902, Bern): "My beloved treasure! Poor dear sweetheart, what you have to suffer that you now can't even write to me yourself. It is such a shame that our dear Lieserl must be introduced to the world this way! I hope you are bright and cheerful by the time my letter arrives.... Now you see that it really turned out to be a Lieserl, just as you had wished. Is she healthy? Does she cry properly? What are her eyes like? Which one of us does she resemble more? Who is giving her milk? Is she hungry?... Couldn't you have a photograph made of her when you are again in good health? Can she turn her eyes in different directions to see things? You can now make observations. I'd like to have a Lieserl myself some day—it must be fascinating. She is certainly able to cry already, but won't learn how to laugh until much later. Therein lies a profound truth. When you feel a little better you will have to draw a picture of her." He has placed an advertisement in the local newspaper, offering to give private lessons.

"What Has Befallen Lieserl"

Now briefly the rest of the story of this star-crossed couple's early years The marriage took place on 6 January 1903. Lieserl, whose existence was unknown to all Einstein scholars until these letters were found and published in 1987, is mentioned in only one other letter of Albert, on 19 September 1903, when Mileva is back at her family's home, as he puts it, "to hatch a new chick"—that is, to give birth to the first of their two sons, Hans Albert. Einstein writes, "I am very sorry about what has befallen Lieserl. It is so easy to suffer lasting effects from scarlet fever. If only this will pass safely. As what is the child registered? We must be very careful that problems don't arise for her later." The Einstein Collected Papers Project mounted an extensive search to trace her through birth registers and even gravestones; but nothing has been found. If she survived, very likely she was adopted by one of Mileva's relatives, in accord with custom in those days for such cases. Although now quite improbable, it is of course technically still possible that she might surface, saying, "My name is Anastasia, and I want the family papers."

This is not just idle speculation. Something like this nearly happened. In the 1930s, after Einstein had moved to America, a woman turned up in Germany who claimed to be a long-lost daughter of Einstein. Her story persuaded a number of Einstein's friends in Europe, who conveyed their surprise and dismay to him. Eventually it turned out that she was a fraud, an actress born in Vienna in 1894, when Einstein was 15. That episode illustrates one of the many cases in which the mature Einstein, like other highly visible persons, became a target of con artists and sensationalizers. That, in part, explains why in Einstein's later years those near him tended to seem so overprotective.

But there is no doubt that the birth of the real Lieserl took a heavy toll on Mileva herself, whose attitudes and behavior tended to become more and more brooding after those secret events. Perhaps she was blaming herself for the decision to give up her daughter, or blamed Albert for acquiescing in it, or possibly even for urging her to acquiesce. We simply do not know. At any rate, the periods of melancholy that increasingly enveloped Mileva became in time evident to everyone. Einstein later described her as taciturn, suspicious, and depressed, and attributed it to a background of schizophrenia on her mother's side. Albert and Mileva's younger son, Eduard, also showed signs of the illness. More and more, the dark side of life began to show itself to that family.

But at least the first few years of marriage seem to have been pleasant enough, on the whole. Shortly after the wedding Albert wrote to his good friend Michele Besso (January 1903): "Well, now I am a thoroughly married husband and lead a nice comfortable life with my wife. She takes excellent care of everything, cooks well, and is always in a good mood." Similarly, Mileva wrote on 20 March 1903 to her confidante, Helene Savić: "I am, if possible, even more attached to my dear treasure than I was in the Zurich days. He is my only companion in society, and I am happiest when he is beside me."

Baby Carriage and Cheap Cigars

A family friend has described life after the birth of Hans Albert in May 1904, during Einstein's most productive phase in his Swiss years. Mileva is out on some errand, and Albert is sitting at the kitchen table, one hand pushing the baby carriage back and forth, while the other is writing one of his world-shaking papers. The whole room is enveloped in the acrid smoke from one of the cheap cigars which were then his luxury. But what is ominously missing from all these comfortable images of the early years of marriage is any evidence of the kind of life they planned when first drawn to each other—two school teachers coming home after work for an evening of studying together happily, and perhaps writing something new for the major physics journals, as a few such teachers then still could hope to do.

The rest is also not what they had dreamt of in their early letters. By 1909, Einstein's reputation is rising fast in science; but the marriage is clearly in trouble, as seen by letters to friends. She writes (again to Savić): "You see, with such fame [for Einstein] not much time remains for his wife.... The pearls are given to the one, the other gets just the shell.... You see, I am very starved for love." All traces of her interest in a career in science or in teaching are gone. Descriptions of her by friends during those years are full of phrases such as brooding, dark moods, introversion, long periods of silence even at home, extreme taciturnity, unresponsiveness, melancholia. The happy *Jugendliebe* between Albert and Mileva had ended. In 1911, the family moved to Prague while Einstein served for about a year as a university professor, but where everyone in that family felt uncomfortable. In 1912, they returned to Zurich, where he had been offered a professorship at their old school, the Poly.

Shortly before World War I, Albert accepted the call to the distinguished position as member of the Prussian Academy of Sciences and research professor at the Kaiser Wilhelm Institute in Berlin. For him, living in the country he had chosen to leave as a youngster was hard enough, but Mileva found it unbearable. She packed up the two boys and took them back to Switzerland in 1914. During the following few years, while Einstein persevered through bouts of illness in almost superhuman effort to fashion the General Relativity Theory, he was also courted by a cousin in Berlin, Elsa Löwenthal, with whom he had an occasional flirtation starting in 1912. In 1919, Mileva divorced Albert and continued to care for her children as best she could. Hans Albert went on to become a professor of engineering at the University of California, Berkeley. But the younger son, Eduard, was in and out of psychiatric clinics, treated for schizophrenia, and he died in 1965 in the same Swiss clinic in which his mother's sister Zorka had also been treated for mental instability.[4] Soon after the divorce, Elsa Löwenthal and Albert were married. When Einstein received the Nobel prize in 1922—which knowledgeable physicists had predicted would come sooner or later—he transferred the Nobel prize money to Mileva, as he had promised in the divorce proceedings. In 1948, in poor physical and mental condition, Mileva died in Switzerland.

What Have We Learned?

So, what have we learned? Some main points would include these:

1. Thanks to these letters, we know of course much more about Mileva Marić, her courage, hopes, early ambition and disappointments. She left no evidence of originality as a future major scientist—but that is of course

true also for most who aspire to such a career, whether male or female. During the early, good years she and Albert longed palpably for each other's companionship. Einstein also valued her intellectually, and not only because he, an autodidact throughout his life, always needed someone who understood him, to talk with about new ideas (as did Niels Bohr). One remembers the warning the physicist Max von Laue gave a friend in 1912 in preparing him for his first meeting with Einstein: "You should be careful that Einstein doesn't talk you to death. He loves to do that, you know."

We also know more about Mileva's increasing demoralization or loss of self-esteem—caused perhaps by the loss of Lieserl, or by the increasing fame of her husband with all the demands this entailed, or by their disappointment with each other, when the original project of a cozy couple studying science together came to naught.

2. And we know more about the young Albert Einstein. In the years I have covered, he was of course not the sad-eyed icon and patriarch of late years, but a lusty and irrepressible young man who attempted to combine his emotional and intellectual passions. This is symbolized for me by the fact that some of his early letters to Mileva were written on pages torn from his notebooks, the back of the sheet containing calculations from the theory of electromagnetism. I recall here the answer Erik Erikson said he received from Freud when Erikson asked what might be the primary components of a good life. Freud had put it memorably in three words: *Arbeit und Liebe* (work and love).

We already knew that in his later years Einstein was never discouraged by failures in his doomed, late project to find a unified field theory; but here we also saw that from early on practically nothing could squelch Einstein's fundamental optimism, either about his work in science or about his faith in Mileva.

3. As I noted, during the years when Einstein was passionately attached to Mileva Marić, and chiefly when she seemed in need of psychological support, he occasionally used in his letters to her such phrases as "our work" or "our investigation." Starting in 1990, a small number of writers attempted to inflate the potential meaning of these phrases to include the strong possibility that Mileva was responsible for either the physics or the mathematics in Einstein's published relativity paper of 1905, but was deprived of any credit for it. For a time it caused extensive and sensationalistic press coverage around the world, even though one of its two main promoters revealed his purpose in an interview (*New York Times*, 27 March 1990): "My point is to say that the king had no clothes." But careful analysis of the matter by established scholars in the history of physics, including John Stachel, Jürgen Renn, Robert Schulmann and Abraham

Pais, has shown that scientific collaboration between the couple was minimal and one-sided. Einstein's occasional use of the word "our" was chiefly meant to serve the emotional needs of the moment. To quote Stachel's essay[1]: "In summary, Marić seems to have encouraged and helped Einstein in a number of ways during their years together, notably by playing the necessary role of alter ego, to whom Einstein could express himself freely while developing his ideas. She also may have helped him by looking up data, suggesting proofs and/or checking calculations and copying some of his notes. He never publicly acknowledged this help [nor did she claim it in her letters to him or to anyone else], nor did a truly creative collaboration ever develop."

Ironically, the exaggeration of Mileva's scientific role, far beyond what she herself ever claimed or could be proved, only detracts from her real and significant place in history. For she was one of the pioneers in the movement to bring women into science, even if she did not reap its benefits. At great personal sacrifice, as it later turned out, she seems to have been essential to Albert during the onerous years of his most creative early period, not only as an anchor of his emotional life, but also as a sympathetic companion with whom he could sound out his highly unconventional ideas during the years when he was undergoing the quite unexpected, rapid, metamorphosis from eager student to a first-rank scientist.

4. From these letters we now know more of Einstein's steps toward his major works of 1905 and beyond: what he read and when (e.g., Wien), how long it took him to go through the early, formative stages of what were to become the papers on relativity, Brownian movement, and the quantum theory of light.

5. We have also relearned the old lesson that establishing and persisting in a long-term loving relationship was for Albert and Mileva, as it is for so many, even more difficult than working on the awesome scientific problems which Einstein knew how to solve. We can never claim to know what it felt like to be them. But their attempts during the first few years to include each other in their life and their work—to make it one life—might be seen in terms of a process of the loosening and expanding of the boundary of one's ego so as to include another—not far from the Platonic theory of love as the enhancement or expansion of the self when it is completed by merging with its cognate partner. Or, to put it better, in the words of Shakespeare's Portia: "One half of me is yours, the other half yours, Mine own I would say; but if mine, then yours, And so all yours!" In this particular case, the initial expectations of making it *one* life were evidently quite unrealistic, and each side felt it eventually had good reasons for feeling deeply disappointed. At any rate, when Einstein's second

marriage (to Elsa) also proved unsatisfactory, Einstein accused himself wistfully of failing rather ignominiously in two attempts.

6. And finally, we have learned that some mysteries remain more or less as they were before. The surviving letters are like fragments of pottery from which one might guess at an ancient civilization. On the personal side, neither of the two people can be fully comprehended at this distance. On the scientific side, these letters do not help explain how a junior civil servant, with only the formal training of a school teacher, could rearrange the structure of physics set up by such giants as Maxwell, Boltzmann, Lorentz, and Planck. Perhaps (as proposed by Renn and Schulmann) Einstein was helped greatly by approaching physics first through reading, as an adolescent, a popular work on natural science, written in the middle of the nineteenth century by Aaron Bernstein. In his *Autobiographical Notes,* Einstein refers to it as one of the books he had the good fortune of hitting upon, which were "not too particular in their logical rigor, but which made up for this by permitting the main thoughts to stand out clearly and synoptically.... [Particularly Bernstein's set of volumes] limited itself almost throughout to qualitative aspects...a work which I read with breathless attention." In this way he gained early a broad conceptual overview of physics on which to build and add technical details. That overview allowed him to draw connections between pedagogically separate subjects when he encountered them in university courses. In short, Einstein started his formal training with a preference for seeing phenomena in the context of the unity of nature, as drawn from Bernstein's writings and the widely-read *Kosmos* of Alexander von Humboldt.

One might conclude that Einstein began with the outlines of a valid scientific worldview rather than with the usual student's concern to master an encyclopedia of individual subjects. This is one more example of Einstein's position as an outsider, a "marginal" person in science, not obedient to any of the existing theories, and hence free to let his driving curiosity and overflowing originality create new ones—even while he and Mileva Marić, against all the obstacles and prejudices, tried to put their intertwined passions for science and for each other into the service of forging a joined life together.

References

1. John Stachel, "Einstein and Marić: A Failed Collaboration," in H. Pycior, N. Slack, P. Abir-Am, eds., *Creative Couples in Science* (New Brunswick, NJ: Rutgers University Press, 1995). Parts of my presentation are indebted to his work.

2. Some of the relevant work of sociologists and psychologists on this subject has been summarized in Gerald Holton, *The Scientific Imagination: Case Studies,* chapter 7, "On the psychology of scientists, and their social concerns" (Cambridge: Cambridge University Press, 1978).

3. The autobiographical passage quoted earlier is in Albert Einstein, "Autobiographical Notes," pp. 3–4, in Paul A. Schilpp, ed., *Albert Einstein: Philosopher-Scientist* (La Salle, IL: Open Court, 1949).

4. Some journalists have recently sought for the genetic origin of Eduard's mental condition not in Mileva's family but in Albert himself. Their source is a theory of "creativity" promoted by Anthony Storr (*The Dynamics of Creation,* New York: Athenaeum, 1972), who begins by announcing that "creative activity [is] a peculiarly apt way for the schizoid individual to express himself" (p. 57). Then he selects his prime target as follows: "Einstein provides the supreme example of how schizoid detachment can be put to creative use," although generously allowing that "yet he retained sufficient contact with reality for his thought to be scientifically viable" (pp. 61, 63). But Storr warns that the price being paid for the "detachment" needed for making "a new model of the universe" is severe: "Such detachment can only be achieved by the person with a predominantly schizoid psychopathology" (p. 67). From there, Dr. Storr moves on to deal similarly with Isaac Newton.
The whole argument reads like a parody of ancient psychiatric work; but it fits with today's trend to construct revisionist accounts of major innovators.

Bibliographical Sources

Einstein, Elizabeth Roboz, *Hans Albert Einstein: Reminiscences of His Life and Our Life Together* (Iowa City, IA: University of Iowa, 1991).

Erikson, Erik H., "Psychoanalytic Reflections on Einstein's Centenary," in G. Holton and Y. Elkana, eds., *Albert Einstein: Historical and Cultural Perspectives* (Princeton, NJ: Princeton University Press, 1982).

Frank, Philipp, *Einstein. His Life and Times* (New York: Alfred A. Knopf, 1947).

Michelmore, Peter, *Einstein. Profile of the Man* (New York: Dodd, Mead & Co., 1962).

Pais, Abraham, *Einstein Lived Here* (New York: Oxford University Press, 1994).

Reiser, Anton, *Albert Einstein* (New York: Albert & Charles Boni, 1930).

Renn, Jürgen and Robert Schulmann, eds., *Albert Einstein, Mileva Marić: The Love Letters* (transl., Shawn Smith)(Princeton, NJ: Princeton University Press, 1992).

Stachel, John, ed., *The Collected Papers of Albert Einstein, Vol. I: The Early Years, 1879–1902* (Princeton, NJ: Princeton University Press, 1987).

Stachel, John, "Einstein and Marić: A Failed Collaboration" [to appear in H. Pycior, N. Slack, P. Abir-Am, eds., *Creative Couples in Science* (New Brunswick, NJ: Rutgers University Press, 1995)].

Trbuhović-Gjurić, Desanka, *Im Schatten Albert Einsteins* (Bern and Stuttgart: Verlag Paul Haupt, 1988) [but see the corrections of the claims, by Pais and Stachel above].

Note: The translations from the German originals of the letters are partly based on existing ones, but are largely the author's. I gladly acknowledge permission from the Einstein Estate to quote from the letters.

What, Precisely, Is "Thinking"?...Einstein's Answer

How did Albert Einstein do his thinking? This is a question often asked by students. And at first glance an answer seems impossible. His work was carried out at the very frontiers of physics and of human ability. And his mind was not open to easy study from the outside, even by those who worked with him—as was discovered by the physicist Banesh Hoffmann who, with Leopold Infeld, was Einstein's assistant in 1937. Hoffmann has given an account of what it was like when he and Infeld, having come to an impassable obstacle in their work, would seek out Einstein to get help. At such a point, Hoffmann related,

> We would all pause and Einstein would stand up quietly and say, in his quaint English, "I will a little think." So saying, he would pace up and down and walk around in circles, all the time twirling a lock of his long grey hair around his forefinger. At these moments of high drama, Infeld and I would remain completely still, not daring to move or make a sound, lest we interrupt his train of thought.

Many minutes would pass this way, and then, all of a sudden,

> Einstein would visibly relax and a smile would light up his face...then he would tell us the solution to the problem, and almost always the solution worked....The solution sometimes was so simple we could have kicked ourselves for not having been able to think of it by ourselves. But that magic was performed invisibly in the recesses of Einstein's mind, by a process that we could not fathom. From this point of view the whole thing was completely frustrating.[1]

But if not accessible from the outside, Einstein's mind was accessible from the inside, because like many of the best scientists, he was interested in the way the scientific imagination works and wrote about it frankly. As far as possible, we shall follow the description, quite accessible to students and in his own words, of how he wrestled with theories of fundamental importance. Needless to say, we shall not be under any illusion that by doing so we can imitate or even fully "explain" his detailed thought processes, nor will we forget that other scientists have other styles.

There are numerous sources to draw on, for Einstein discussed his view of the nature of scientific discovery, in a generally consistent way, on many occasions, notably in the essays collected in the book *Ideas and Opinions* and in his letters. He was also intrigued enough by this problem to discuss it with a number of researchers into the psychology of scientific ideas, including Max Wertheimer, Jacques Hadamard, and Jean Piaget, and with several philosophers of science. Indeed, from his earliest student days, Einstein was deeply interested in the theory of knowledge (epistemology). He wrote, "The reciprocal relationship of epistemology and science is of noteworthy kind. They are dependent upon each other. Epistemology without contact with science becomes an empty scheme. Science without epistemology is—insofar as it is thinkable at all—primitive and muddled."[2]

There are two especially suitable routes to Einstein's thoughts. One is a set of pages near the beginning of an essay, "Autobiographical Notes," which he wrote in 1946 as the opening article for the volume *Albert Einstein: Philosopher-Scientist*, edited by Paul A. Schilpp.[3] It is the only serious autobiographical essay he ever wrote, and he called it jokingly his own "obituary." It gives a fascinating picture of Einstein's contributions as he viewed them, looking back at the age of 67. The essay is chiefly an account of his intellectual development rather than an autobiography in the usual sense. We shall now use this remarkable document to learn from his own words, while avoiding the use of technical, philosophical terminology, as he himself avoided it. All quotations not otherwise identified are from the first thirteen pages of his autobiography. It may be instructive to open to these pages and follow them in parallel with the analysis given below, which also draws on his other writings.

The other path to an understanding of Einstein's way of thinking is found in some letters he wrote to an old friend after publication of the "Autobiographical Notes." These allow Einstein to rebut, so to speak, a few of the objections a reader of the autobiographical essay might have, and I discuss them therefore at the end of this chapter, where Einstein should have the final word.

The Courage to Think

It certainly is curious to start one's autobiography, not with where and when one was born, the names of one's parents, and similar personal details, but to focus instead on a question which Einstein phrases simply: "What, precisely, is 'thinking'?" Einstein explains[4] why he has to start his "obituary" in this way: "For the essential in the being of a man of my type lies precisely in *what* he thinks and *how* he thinks, not in what he does or suffers.[5]

From this viewpoint, thinking is not a joy or a chore added to the daily existence. It is the essence of a person's very being, and the tool by which the transient sorrows, the primitive forms of feeling, and what he calls the other "merely personal" parts of existence can be mastered. For it is through such thought that man can lift himself up to a level where he can think about "great, eternal riddles." It is a "liberation" which can yield inner freedom and security. When the mind grasps the "extra-personal" part of the world—that part which is not tied to shifting desires and moods—it gains knowledge which all men and women can share regardless of individual conditions, customs, and other differences.

This, of course, is precisely why the laws of nature, towards which these thoughts are directed, are so powerful: their applicability can be demonstrated by anyone, anywhere, at any time. The laws of nature are utterly shareable. Insofar as the conclusions are right, the laws discovered by a scientist are equally valid for different thinkers, or *invariant* with respect to the individual personal situations. Einstein's interest in this matter seems to be not unrelated to his work in the physics of relativity: the essence of relativity theory is precisely that it provides a tool for expressing the laws of nature in such a manner that they are invariant with respect to differently moving observers.

As his "Autobiographical Notes" show, Einstein was also aware that life cannot be all thought, that even the enjoyment of thought can be carried to a point where it may be "at the cost of other sides" of one's personality. But the danger which more ordinary persons face is not that they will abandon their very necessary personal ties, but that the society surrounding them will not say often enough what Einstein here suggests to his wide audience: that the purpose of thinking is more than merely solving problems and puzzles. It is instead, and most importantly, the necessary tool for permitting one's intellectual talent to come through, so that "gradually the major interest disengages itself...from the momentary and merely personal." Here Einstein is saying: Have the courage to take your own thoughts seriously, for they will shape you.

Thinking with Images

Having touched on the *why* of thinking, the autobiography takes up the *how* of thinking and strangely seems to be concerned with "pictures" (*Bilder*):

> What, precisely, is "thinking"? When, at the reception of sense-impressions, memory-pictures emerge, this is not yet "thinking." And when such pictures form series, each member of which calls forth another, this too is not yet "thinking." When, however, a certain picture turns up in many such series, then—precisely through such return—it becomes an ordering element for such series, in that it connects series which in themselves are unconnected. Such an element becomes an instrument, a concept.[6]

Adhering to one of several contesting traditions in psychology and philosophy and perhaps particularly influenced by Helmholtz and Boltzmann, Einstein holds that the repeated encounter with images (such as "memory pictures") in a different context leads to the formation of "concepts." Thus, a small child might form the concept "glass" when he or she experiences that a variety of differently shaped solids are hard, transparent and break on being dropped.

A concept must of course eventually be put into a form where it can be communicated to others; but for private thought it is not necessary to wait for this stage. For some people, including such physicists as Faraday and Rutherford, the most important part of thinking may occur without the use of words. Einstein writes: "I have no doubts but that our thinking goes on for the most part without use of signs (words) and beyond that to a considerable degree unconsciously."[7] Such persons tend to think in terms of images to which words may or may not be assignable; Einstein tells of his pleasure in discovering, as a boy, his skill in contemplating relationships among geometrical "objects"—triangles and other nonverbal elements of the imagination. He explained more fully in a letter to Jacques Hadamard:

> The psychical entities which seem to serve as elements in thought are certain signs and more or less clear images which can be "voluntarily" reproduced and combined.... But taken from a psychological viewpoint, this combinatory play seems to be the essential feature in productive thought— before there is any connection in words or other kinds of signs which can be communicated to others. The above-mentioned elements are, in my case, of visual and some muscular type. Conventional words or other signs have to be sought for laboriously only in a secondary stage, when the mentioned associative play is sufficiently established and can be reproduced at will.[8]

Gedankenexperimente

Einstein's ability to visualize may have helped him in the brilliant use he made of "thought experiments" (*Gedankenexperimente*). His first came to him at the age of about 16, when he tried to imagine that he was pursuing a beam of light and wondered what the observable values of the electric and magnetic field vectors would be in the electromagnetic wave making up the light beam.[9] For example, looking back along the beam over the space of one whole wavelength, one should see that the local magnitudes of the electric and magnetic field vectors increase point by point from, say, zero to full strength, and then decrease again to zero, one wavelength away. This seemed to him a paradoxical conclusion. Already at that age, he seems to have assumed that Maxwell's equations must remain unchanged in form for the observer moving along the beam; but from those equations one did not expect to find such a spatially oscillatory pattern of **E** and **B** vectors in free space. He realized later in this problem "the germ of the special relativity theory was contained."

Among other examples of visualized *Gedankenexperimente*, Einstein related one which he said had led him to the general theory of relativity. In 1907 he was trying to modify the special theory of relativity; it applied to systems moving with constant velocity, and so did not accommodate in a natural way the behavior of accelerating objects, hence was not applicable to gravitation.

> At that point there came to me the happiest thought of my life [sic!], in the following form: Just as in the case with the electric field produced by electromagnetic induction, the gravitational field has similarly only a relative existence. *For if one considers an observer in free fall, e.g., from the roof of a house, there exists for him during his fall no gravitational field*—at least in his immediate vicinity. For if the observer releases any objects they will remain relative to him in a state of rest, or in a state of uniform motion, independent of their particular chemical and physical nature. (In this consideration one must naturally neglect air resistance.) The observer therefore is justified to consider his state as one of "rest."

> The extraordinarily curious, empirical law that all bodies in the same gravitational field fall with the same acceleration received through this consideration at once a deep physical meaning. For if there is even a single thing which falls differently in a gravitational field than do the others, the observer would discern by means of it that he is in a gravitational field, and that he is falling in it. But if such a thing does not exist—as experience has confirmed with great precision—the observer lacks any objective ground to consider himself as falling in a gravitational field. Rather, he has the right to consider his state as that of rest, and his surroundings (with respect to gravitation) as fieldfree.

The fact of experience concerning the independence of acceleration in free fall with respect to the material is therefore a mighty argument that the postulate of relativity is to be extended to coordinate systems that move non-uniformly relative to one another....[10]

The Free Play with Concepts

Having stressed the role of images and memory pictures, including *Gedankenexperimente*, in thinking, and having defined "concepts" as the crystallized products, the unvarying elements found to be common to many series of such memory pictures, Einstein now makes a startling assertion: "All our thinking is of the nature of a free play with concepts." This sentence has to be unraveled for it deals with the eternal antithesis between the two indispensable elements in all human thought, the empirical and the rational.

Even if one grants that "free play" is still play within definite, if only temporarily chosen, rules—similar to tentatively trying out a word to see if it fits into a crossword puzzle—by no means all philosophers would agree with Einstein's position. Some would argue that the external world imposes itself strongly on us and gives us little leeway for play, let alone for choosing the rules of the game. In Einstein's youth, most of his contemporaries believed in Immanuel Kant's description of the boundaries of such "play," namely, that they were fixed by two intuitions which are present in one's mind already at birth (i.e., *a priori*): Newtonian absolute space and absolute time. Only a few disagreed, including Ernst Mach, who called absolute space "a conceptual monstrosity, purely a thought thing which cannot be pointed to in experience."

Thus Einstein was struggling anew with the old question: What precisely is the relation between our knowledge and the sensory raw material, "the only source of our knowledge?"[11] If we could be sure that there is one unchanging, external, "objective" world that is connected to our brains and our sensations in a reliable, causal way, then pure thought can lead to truths about physical science. But since we cannot be certain of this, how can we avoid falling constantly into error or fantasy? David Hume had shown that "habit may lead us to belief and expectation but not to the knowledge, and still less to the understanding, of lawful relations."[12] Einstein concluded that, "In error are those theorists who believe that theory comes inductively from experience."[13]

In fact, he was skeptical about both of the major opposing philosophies. He wrote that there is an "aristocratic illusion [of subjectivism or idealism] concerning the unlimited penetrating power of thought," just as there is a "plebian illusion of naive realism, according to which things are as they are perceived by us through our senses."[14] Like Ernst Mach[15], Einstein held that there is no "real world" to which one can repair—the whole concept of the "real world"[16] being justified only insofar as it refers to the mental connections that weave the multitude of sense impressions into some connected net. Sense impressions are "conditioned by an 'objective' and by a 'subjective' factor."[17] Similarly, reality itself is a relation between what is outside us and inside us. "The real world is not given to us, but put to us (by way of a riddle)."[18]

Since the world as dealt with by a scientist is more complex than was allowed for in the current philosophies, Einstein thought that the way to escape illusion was by avoiding being a captive of any one school of philosophy. He would take from any system the portions he found useful. Such a scientist, he realized "therefore must appear to the systematic epistemologist as a type of unscrupulous opportunist: he appears as a *realist* insofar as he seeks to describe the world independent of the acts of perception; as *idealist* insofar as he looks upon the concepts and theories as the free inventions of the human spirit (not logically derivable from what is empirically given); as *positivist* insofar as he considers his concepts and theories justified *only* to the extent to which they furnish a logical representation of relations among sensory experiences. He may even appear as *Platonist* or *Pythagorean* insofar as he considers the viewpoint of logical simplicity as an indispensable and effective tool of his research."[19]

But what justifies this "free play with concepts?" There is only one justification: that it can result, perhaps after much labor, in a thought structure which gives us the testable realization of having achieved meaningful order over a large range of sense experiences that would otherwise seem separate and unconnected. In the important essay "Physics and Reality,"[20] which covers much the same ground as the early pages of the "Autobiographical Notes," Einstein makes the same point with this fine image: "By means of such concepts and mental relations between them we are able to orient ourselves in the labyrinth of sense impressions."[21]

This important process is described by Einstein in a condensed paragraph of the "Autobiographical Notes."[22] "Imagine," he says, "on one side the totality of sense experiences," such as the observation that the needle on a meter is seen to deflect. On the other side, he puts the "totality of concepts and propositions which are laid down in books," which comprises the distilled products of past progress such as the concepts of force or momentum, propositions or axioms that make use of such concepts (for example, the law of conservation of momentum), and more generally, any concepts of ordinary thinking (for example, "black" and "raven"). Investigating the relations that exist among the concepts and propositions is "the business of logical thinking," which is carried out along the "firmly laid-down rules" of logic. The rules of logic, like the concepts themselves, are of course not God-given but are the "creation of man." However, once they are agreed upon and are part of a widely-held convention—the rules of syllogism, for example—they tell us with (only seemingly) inescapable finality that if all ravens are black and a particular bird is a raven, then the bird is black. They allow us to deduce from the law of conservation of momentum that in a closed system containing only a neutron and proton, the momentum gained by one is accompanied by the loss experienced by the other. Without the use of logic to draw conclusions, no disciplined thinking, and hence no science, could exist.

But all such conclusions, Einstein warns, are empty of useful "meaning" or "content" until there is some definition by which the particular concept (e.g., "raven" or "neutron") is correlated with actual instances of the concept which have consequences in the world of experience rather than in the world of words and logical rules. Necessary though the correlation or connection between concepts and sense experience is, Einstein warns that it is "not itself of a logical nature." It is an act in which, Einstein holds, "intuition" is one guide, even if not an infallible one. Without it, one could not be led to the assertion that a particular bird, despite some differences in its exact size or degree of blackness from all other birds, does belong to the species raven; or that the start of a particular track, visible in the cloud chamber, is the place where a neutron has struck a proton.

One might wish that Einstein had used a notion more firm than the dangerous-sounding one of "intuition." But he saw no other way. He rejected the use of the word "abstraction" to characterize the transition from observation to concept, e.g., from individual black birds to the idea of "raven." He rejected it precisely because, he said, "I do not consider it justifiable to veil the logical independence of the concept from the sense experiences" (whereas the use of the term abstraction or induction might make it seem as if there *were* a logical dependence).

The danger is evidently that delusion or fantasy can and does make similar use of the elements of thinking: and since there are no hard, utterly reliable connections between the concepts, propositions, and experience, one cannot know with absolute certainty whether one has escaped the trap of false conclusion. That is why it was thought for so long that observations proved the sun went around the earth; that time had a universal meaning, the same for all moving observers; and that Euclidean geometry is the only one that has a place in the physical world. But this is just where Einstein's view is most helpful: only those who think they *can* play freely with concepts can pull themselves out of such error. His message is even more liberal: the concepts themselves, in our thoughts and verbal expressions are, "when viewed logically, the free creation of thought which cannot inductively be gained from sense experience."[23] We must be continually aware that it is not necessity but habit which leads us to identify certain concepts (for example, "bread") with corresponding sense experience (feel, smell, taste, satisfaction); for, since this works well enough most of the time, "we do not become conscious of the gulf—logically unbridgeable—which separates the world of sense experience from the world of concepts and propositions." Einstein is perhaps so insistent on the point because he had to discover it the hard way: as a young man, he had to overcome the accepted meanings of such concepts as space, time, simultaneity, energy, etc., and to propose redefinitions that reshaped all our physics, and hence our very concept of reality itself. One might well add here that Einstein demanded the same freedom to challenge orthodoxy outside science. Thus, as a boy he rejected the malignant militarism he saw entrenched in the life of his native country.

Once a conceptual structure has tentatively been erected, how can one check whether it is scientifically "true?" It depends on how nearly the aim of making the system deal with a large amount of (ideally, cover the totality of) diverse sense experience has been achieved, and how economical or parsimonious the introduction of separate basic concepts or axioms into system has been. Einstein doubted a physical theory, and would say that it failed to "go to the heart of the matter," if it had to be jerry-built with the aid of *ad hoc* hypotheses, each specially introduced to produce greater agreement between theory and experience (experiment). He also was rarely convinced by theories that dealt with only a small part of the range of physical phenomena, applicable only here or there under special circumstances. In this view, a really good theory, one that has high scientific "truth" value, is considered to be correct not merely when it does not harbor any logical contradictions, but when it allows a close check on the correspondence between the predictions of the theory and a large range of possible experimental experiences. He summarized all this in the following way: "One comes nearer to the most superior scientific goal, to embrace a maximum of experimental content through logical deduction from a minimum of hypotheses....One must allow the theoretician his imagination, for there is no other possible way for reaching the goal. In any case, it is not an aimless imagination but a search for the logically simplest possibilities and their consequences."[24]

This search may take "years of groping in the dark;" hence the ability to hold onto a problem for a long time, and not to be destroyed by repeated failure, is necessary for any serious researcher. As Einstein once said, "Now I know why there are so many people who love chopping wood. In that activity one immediately sees the results." But for him, the goal of "embracing a maximum of experimental content...with a minimum of hypotheses" meant nothing less than endless devotion to the simplification and unification of our world picture by producing fusions in hitherto separate fundamental concepts such as space and time, mass and energy, gravitation and inertial mass, electric and magnetic fields, and inertial and accelerating systems.

Keeping Alive the Sense of Wonder

Before further unraveling Einstein's views on how to think scientifically about the deep problems, let us consider, at least briefly, an engaging passage in the "Autobiographical Notes" in which Einstein speaks of the importance of the sense of marvel, of deep curiosity, of "wonder" such as his two experiences, when, at the age of 4 or 5, he was shown a magnetic compass by his father, and when, at the age of 12, a book on Euclidean geometry came into his hands.[25] A person's thought-world develops in part just by mastering certain new experiences which were so inexplicable, in terms of the previous stage of development, that a sense of wonder or enchantment was aroused. As we learn more, both through science and other approaches, we progressively find that the world around us, as it becomes more rational, also becomes more "disenchanted." But Einstein repeatedly insisted in other writings that there is a limit to this progressive disenchantment, and even the best scientist must not be so insensitive or falsely proud as to forget it. For, as Einstein said in a famous paragraph: "It is a fact that the totality of sense experiences is so constituted as to permit putting them in order by means of thinking—a fact which can only leave us astonished, but which we shall never comprehend. One can say: The eternally incomprehensible thing about the world is its comprehensibility."

He went on: "In speaking here of 'comprehensibility,' the expression is used in its most modest sense. It implies: the production of some sort of order among sense impressions, this order being produced by the creation of general concepts, by relations among these concepts, and by relations of some kind between the concepts and sense experience. It is in this sense that the world of our sense experiences is comprehensible. The fact that it is comprehensible is a wonder."[26]

That wonder [*Wunder*], that sense of awe, can only grow stronger, Einstein implied, the more successfully our scientific thoughts find order to exist among the separate phenomena of nature. This success aroused in him a "deep conviction of the rationality of the universe;" to this conviction he gave the name "cosmic religious feeling," and he saw it as the "strongest and noblest motive for scientific research."[27] Indeed, "the most beautiful experience we can have is the mysterious. It is the fundamental emotion which stands at the cradle of true art and true science. Whoever can no longer wonder, no longer marvel, is as good as dead..."[28]

After the publication of such sentiments, Einstein received a plaintive letter from one of his oldest and best friends, Maurice Solovine. They had met in Bern in 1902 when Einstein was 23 years old, and they became close friends. Solovine was then a young philosophy student at the University of Bern, to which he had come from Rumania, and, together with Conrad Habicht, who was also a student at the university, they banded together to meet regularly to read and discuss works in science and philosophy. With high irony they called themselves the "Olympia Academy." Their "dinners" were no banquets: they all lived on the edge of poverty, and Solovine tells us that their idea of a special dinner was two hard-boiled eggs each. But the talk was that much better, as they discussed works by Ernst Mach, J. S. Mill, David Hume, Plato, Henri Poincaré, Karl Pearson, Spinoza, Hermann Helmholtz, Ampère—and also those of Sophocles, Racine, and Dickens. Many of Einstein's epistemological ideas might be traced back to these discussions.

Now, half a century later, Maurice Solovine was worried. He asked Einstein how there could be a puzzle about the understandability of our world. For us it is simply an undeniable necessity, which lies in our very nature. No doubt Solovine was bothered that Einstein's remarks seemed to allow into science, that most rational activity of mankind, a function for the human mind which is not rational in the sense of being coldly logical. But Einstein rejected as a "malady"[29] the kind of accusation that implied that he was become "metaphysical." Instead, he saw the opportunity of using *all* one's faculties and skills to do science as a sign of strength rather than of weakness.

Certainly, he did not propose to abandon rationality, nor to guess where one must puzzle things out in a careful, logical way. But he saw that there is, and has to be, a role for those other elements of thinking which, properly used, can help scientific thought. Specifically, this could become necessary at two points in Einstein's scheme. One is the courageous use of an intuitive feeling for nature *when there is simply no other guide at all*—as when one has tentatively to propose an axiom that by definition is unproved (as Einstein did at the start of the first paper on relativity, where he simply proposed the principle of relativity and the principle of constancy of light velocity); or when one decides which sense experiences to select in order to make an operational definition of a concept. The other point is the sense of wonder at being able to discern something of the grand design of the world, a feeling that motivates and sustains many a scientist.

Einstein's reply (30 March 1952) to Solovine addresses this second point.

> You find it remarkable that the comprehensibility of the world (insofar as we are justified to speak of such a comprehensibility) seems to me a wonder or eternal secret. Now, *a priori*, one should, after all, expect a chaotic world that is in no way graspable through thinking. One could (even *should*) expect that the world turns out to be lawful only insofar as we make an ordering intervention. It would be a kind of ordering like putting into alphabetic order the words of a language. On the other hand, the kind of order which, for example, was created through [the discovery of] Newton's theory of gravitation is of quite a different character. Even if the axioms of the theory are put forward by human agents, the success of such an enterprise does suppose a high degree of order in the objective world, which one had no justification whatever to expect *a priori*. Here lies the sense of "wonder" which increases even more with the development of our knowledge.

> And here lies the weak point for the positivists and the professional atheists, who are feeling happy through the consciousness of having successfully made the world not only God-free, but even "wonder-free." The nice thing is that we must be content with the acknowledgment of the "wonder," without there being a legitimate way beyond it. I feel I must add this explicitly, so you wouldn't think that I—weakened by age—have become a victim of the clergy.

Epilogue

In one of Einstein's other letters to Maurice Solovine, Einstein goes over some of these questions—but this time with the aid of a diagram, as befits a person who prefers to think visually.[30] In this and all these writings, Einstein asks his reader to take the business of making progress in science into one's own hands; to insist on thinking one's own thoughts even if they are not blessed by consent from the crowd; to challenge the presumed inevitability or orthodoxy of ideas that do not meet the test of an original mind; and to live and think in all three portions of our rich world—at the level of everyday experience, the level of scientific reasoning, and the level of deeply felt wonder.

To follow these directives with superb results is not what anyone can aspire to who lacks what Einstein called simply a "scientific instinct."[31] Yet one cannot help but be moved by this message, to think and act with courage, independence, and imagination.

References

1. Quoted from a report by Banesh Hoffmann in G.J. Whitrow, *Einstein, the Man and His Achievement* (New York: Dover, 1967), p. 75.
2. Paul Arthur Schilpp, ed., *Albert Einstein: Philosopher-Scientist* (Evanston, Ill.: Library of Living Philosophers, Inc., 1949), pp. 683–684.
3. Albert Einstein, "Autobiographical Notes" (in German and English), in Schilpp, pp. 1–95.
4. Ibid., p. 7.
5. Ibid., p. 33
6 Ibid., p. 7.
7. Ibid., p. 9
8. Albert Einstein, *Ideas and Opinions*, new translations and revisions by Sonja Bergmann (New York: Crown, 1954), pp. 25–26.
9. Einstein, "Autobiographical Notes," p. 53.
10. From an unpublished manuscript, quoted in G. Holton, "Notes Toward the Psychobiographical Study of Scientific Genius," in Y. Elkana, ed., *The Interaction Between Science and Philosophy* (New York: Humanities Press, 1975), pp. 370–371.
11. Einstein, *Ideas and Opinions*, p. 22
12. Ibid.
13. Ibid., p. 301.
14. Ibid., p. 20.
15. See Gerald Holton, *Thematic Origins of Scientific Thought: Kepler to Einstein* (Cambridge: Harvard University Press, 1988 edition), p. 258.
16. Einstein, *Ideas and Opinions*, p. 291.
17. Schilpp, p. 673.
18. Ibid., p. 680.
19. Ibid., p. 684.
20. Einstein, *Ideas and Opinions*, p. 290 ff.
21. Ibid., p. 291.
22. Einstein, "Autobiographical Notes," pp. 11–13.
23. Ibid., p. 22
24. Ibid., p. 282, in translation by Seelig.
25. Einstein, "Autobiographical Notes," pp. 9–11.
26. Einstein, *Ideas and Opinions*, p. 292 (but in corrected translation).
27. Ibid., p. 39.
28. Ibid., p. 11.

29. Ibid., p. 24.
30. The analysis is the subject of an essay, "Einstein's Model for Constructing a Scientific Theory," chapter 2 in G. Holton, *The Advancement of Science, and Its Burdens: The Jefferson Lecture and Other Essays* (Cambridge, U.K.: Cambridge University Press, 1986).
31. Einstein, *Ideas and Opinions*, p. 318.

Part Two:
SCIENCE IN HISTORY

What Place for Science in Our Culture at the "End of the Modern Era"?

Behind every act in the life of a scientist—whether it be the choice of a research program, or interaction with students, the public and the media, or the never-ending search for funding, or advice asked by government officials—there is a hidden factor that in large part determines the outcome. That factor is how society at large regards the place of science in our culture. Most practitioners of science would claim they have little interest or expertise to concern themselves with such a seemingly complex and amorphous problem—at least not until the time comes, as it does periodically, when they begin to notice that their largely unconscious assumptions about the relations of science and the wider polity are being severely challenged.

Such a time has arrived once more. Here and there, intellectuals are waking up to the fact that increasingly such concepts as the "end of the modern era," the "end of progress," and the "end of objectivity," originating from parts of academe, from eloquent popularizers, and even from members of Congress, are making an unquestioned place for themselves in the public mind, with surprisingly little audible opposition from leaders of the scientific establishment. But far from being a passing phase, connected with the usual *fin-de-siècle* preoccupations, the movement—different from the antiscience phenomenon which I have tried to analyze elsewhere[1]—signals the resurgence of an old, recurring rebellion against some of the Enlightenment-based presuppositions of Western civilization, particularly against the claim of science that it can lead to a kind of knowledge which is progressively improvable, in principle universally accessible (i.e., intersubjective), and potentially valuable and civilizing. The impact of the resurgence of this reaction on the life of the scientist, on the public understanding of science generally, and on the legislation of science policy is measurably growing and will become palpable even for the least attentive.

The aim of this essay is to help understand the movement, its main sources, and its driving ambitions. To this end it is well to begin with a survey of some of the chief theorists on the question of what role, if any, science may play in twentieth-century culture, and its effects on key legislators in the U.S. who are now redesigning the direction and conduct of science. In effect one must look back beyond the so-called implicit "contract" forged in the aftermath of World War II between science and society.

That contract, still the dominant myth among the majority of scientists even while it hardly corresponds to reality today, was the result of a more innocent phase, when for a few decades the pursuit of scientific knowledge was widely thought—above all by the scientists themselves—to embody the classical values of Western civilization, starting with the three primary virtues of truth, goodness, and beauty: when science tended to be praised as a central truth-seeking and enlightening process in modern culture—one might call it the Newtonian search for Omniscience; when science was thought to embody the ethos of practical goodness in an imperfect world, both through its largely self-correcting practice of honor in science and through its tendency to lead to applications that might improve the human condition and ward off the enemies of our form of society—a Baconian search for a benign sort of Omnipotence; and when the discovery of beauty in the structure, coherence, simplicity and rationality of the world was thought of as a Keplerian enchantment, the highest reward for the exhausting labor.

Before the Euphoria Ended

The last time the optimistic description just given could have been said to be generally taken for granted, at least in the United States, was the period following the ending of World War II. It was embodied also in the famous Vannevar Bush report, *Science, the Endless Frontier*, of 1945, which became a main driving force of science policy in that country. Because it is such a convenient example of modern post-Enlightenment optimism about the role of science in culture, one which so many scientists tacitly assume to be still operative, it will be illuminating to look at the main thrust of that document.

In November 1944, President Roosevelt requested from Vannevar Bush, the head of the wartime Office of Scientific Research and Develop ment, a report that would outline how, in the postwar world, research in the natural sciences—he called it "the new frontiers of the mind"—could be strengthened and put to service for the nation and humanity. Roosevelt was particularly interested in three results: waging a new "war of science against disease," "discovering and developing scientific talent in Ameri can youth," and designing a new system of vigorous federal support for scientific research in the public and private sectors. Beyond and above those, he argued that science's applications, so useful during the bitter war to preserve the world from fascist dictatorship (with the successes of the Allies' radar, antisubmarine devices and proximity fuses the most striking examples at that time), now could be harnessed to "create a fuller and more fruitful employment, and a fuller and more fruitful life."

Vannevar Bush's detailed response came less than eight months later, the result of a crash program by an impressive brain trust of about forty experts from industry, academe, and government. Roosevelt had died, but with the war's successful end in sight, the American administration proved generally hospitable to the report's ideas. While some of the de tails were too optimistic and others were modified in practice (often to Bush's dismay), his vision, it is generally agreed, set the stage for the de velopment of new institutions for the support of science during the fol lowing decades, and paralleled the generally favorable popular attitudes that were prerequisites for the actions. The base was laid for global lead ership in many branches of basic science. Not until the Vietnam war esca lated was there substantial popular disenchantment both with governmen tal authority, with the widely visible use of sophisticated technology in a hopeless and unpopular war, and by implication with science that presum ably could help give birth to such abuse. It signaled the end of what might be called a rather euphoric phase in the relation of science and society in this century.

The Bush report, as well as the rival proposals by Senator Harley Kil gore, were historic exemplars of the science-based progressivism reigning at the time, which saw science and democracy as natural allies in the serv ice of the ideal empowerment and instruction of the polity as a whole. In this sense, they were part of the American dream as far back as Benjamin Franklin and his fellow statesmen-science amateurs. Vannevar Bush him self hinted as much in the brief preface to his report, taking courage from the fact that, as he put it, "the pioneer spirit is still vigorous within the nation." And to make the connection with the tradition of Condorcet even more explicit, he added a sentence which, while presenting the reigning opinion of a citizen of the mid-1940s, is likely to be rejected today by many who think of themselves as the children of the 1960s and 1970s. He wrote: "Scientific progress is one essential key to our security as a nation, to our better health, to more jobs, to a higher standard of living, and to our cultural progress." One could hear an echo of Thomas Jefferson's formula: "The important truths [are] that knowledge is power, knowledge is safety, knowledge is happiness."

Bush and his contemporaries could hardly have imagined that by the early 1990s those hopes had begun to be rejected, even at the highest levels— that, for example, a key person in the U.S. Congress for science policy could imply (as we shall see in more detail later) that science and technology alone can be held to account for the whole sorry list of failures of decades of misdirected political leadership; he said: "Global leadership in science and technology has not translated into leadership in infant health, life expectancy, rates of literacy, equality of opportunity, productivity of workers, or efficiency of resource consumption. Neither has it overcome failing education systems, decaying cities, environmental degradation, unaffordable health care, and the largest national debt in history."[2] And another highly placed observer, formerly the Director of the National Science Foundation, exulted: "The days of Vannevar Bush are over and gone...the whole world is changing."

The Changing Balance of Sentiments

After this reminder of a mid-century worldview predominant before the generation now in leadership positions came on the scene, we turn from the level of momentary vagaries to come closer to understanding the causal mechanisms responsible for the changes in the place assigned to science at significant stages in the intellectual history of the past hundred years. For if we know the general causes in the variation of the underlying ideology, we shall better understand the changes in policy toward science at a given moment.

Here we must confront at once the question of whether these changes are gradual, and part of an evolutionary development, or are so sudden that, as if in a political revolution, one passes discontinuously from the end of one age to the beginning of another. If the latter is the case, we would now be passing through a rupture of history, with "modern" behind us and "postmodern" right, left, and all before us. While I doubt this is the case—and it certainly is not visible in the *content* of science as against some of the current writings about science today—a fashion in history proper has for some time been trying to discern the arrival of a new age. Periodization, the arranging of the flow of events into clearly separate eras, is a common tool, although applied more wisely from the safe distance of retrospection. That is how we got such schoolbook chapters as "The Age of Reason," or "The Progressive Era in America" around the turn of the century.

A chastening example of that whole genre was provided by the American historian Henry Adams. At the beginning of this century, he had been impressed by the publications of the chemist J. Willard Gibbs of Yale on the phase rule for understanding heterogeneous equilibria. Adams was also fascinated by the strange idea of some physicists of that day that the phase rule can serve, by analogy, as a means for putting into hierarchical order the following sequence: solid, fluid, gas, electricity, ether, and space, as if they formed a sequence of phases. Stimulated by such ideas, Adams believed that thought itself, too, passed in time through different phases, each representing a different period. In his essay of 1909, "The Rule of Phase Applied to History," Adams came to a remarkable conclusion about the imminent passing of modernity: "The future of Thought," he wrote, "and therefore of History, lies in the hands of the physicist, and...the future historian must seek his education in the world of mathematical physics...[If necessary] the physics departments will have to assume the task alone." Henry Adams' conclusion might fairly have been called in its own day a declaration of what the postmodern age would look like. Today's formulation is likely to be exactly the opposite one.

I cite this example—and many others come to mind—to signal my discomfort with trying to divide history into distinct periods. A less rigid and more workable notion is to recognize that at any given time and place, even during a period when a civilization appears to be in a more or less settled state of dynamic equilibrium, there exist simultaneously several competing and conflicting ideologies within the momentary heterogeneous mixture of outlooks. As Leszek Kolakowski noted, "It is certain that modernity is as little modern as are the attacks on modernity.... The clash between the ancient and the modern is probably everlasting and we will never get rid of it, as it expresses the natural tension between structure and evolution, and this tension seems to be biologically rooted; it is, we may believe, an essential characteristic of life."[3]

It is sometimes possible in retrospect to identify one of the competing worldviews as the most dominant one for a longer or shorter period. But what is also likely to occur when followed in real time are two effects. The first is that each of the different competing groups works fervently to raise its own ideology to a position where it would be accepted as the "taste of the time" or the "climate of opinion" which characterizes that particular age and region. The newest and most ambitious one *will also be trying as part of its agenda to delegitimate the claims of its main rivals.* Especially when the previously relatively stable equilibrium begins to crumble, the pandemonium of contrasting voices gets louder. Some partial victors rise to be major claimants above the rest, and one of them may even be generally recognized for a while as the embodiment of the new worldview or "sentiment" of the society. Secondly, in this constant seesaw of changing historic forces, mankind's inherent liability to engage in over-ambition or one-sidedness may infect some of these claimants (not excluding, on occasion, scientists). This is the tendency, as Hegel had warned, toward "the self-infinitization of man," or simply to *yield to excess*—which, in turn can generate in reaction the same sort of excess among the opposing claimants. Recognizing these two facts is, in my view, central for understanding the course of twentieth-century culture.

In this general struggle, from that of Apollo vs. Dionysus in Greece to this day, the specific, more limited question of the place assigned to the scientific conception of the world has always played a part. Sometimes this place has been at the cherished core of the rising or victorious overall worldview; sometimes it has found itself embedded in the sinking or defeated one, and then was even accused of nourishing a great variety of sins against the better interests of humanity.

Historians of ideas have mapped the changing forms of the general contrary trends. Wise political leaders, too, have at times watched with apprehension as the net balance of prevailing sentiments has taken a turn, for as Jefferson said, "It is the manner and spirit of a people which preserve a republic in vigor. A degeneracy in these is a canker which soon eats into the heart of its laws and constitution." Weighty scholarship has chronicled how one of the world conceptions, and the scientific position within it, gained predominance over the others for some decades in significant segments of Western culture—an example is Robert K. Merton's early study on science and seventeenth-century Puritanism. There is also much documentation that such sentiments subsequently gave ground, as the overall balance of benignity or distress moved the other way for some more decades. As to the practicing scientists themselves, most of them have paid little attention to this constant seesaw of sentiments, except to weigh in now and then as promoters of the positive swings, or occasionally to become victims of the negative ones.

Today, at our own *fin-de-siècle*, this oscillating spectacle, so engrossing to the scholar, has ceased to be merely the site for the research of historians. The general balance among the contending elements, and with it the attitude of traditional patrons, is changing before our eyes. Studying this current drama is as fascinating and fruitful for the historian of ideas, whose perspective I shall be taking here, as the appearance of a supernova may be for an astronomer. But in both cases, the present state is the product of an historic process, the latest member of a motley progression.

Toward a "Monistic Century"

Let us therefore look at some of those claimants for representing the climate over the past hundred years up to the present—a sequence of selected samples meant to be analogous to stages in the growth of a culture of cells seen under the microscope. Our first sample concerns an event as the new century was signaling its beginning: the World's Columbia Exposition at Chicago in 1893. The fair was intended as a triumphant celebration of human and social progress in all fields—above all, industrial, scientific, and architectural. The big attractions were Machinery Hall, the Electricity Building, the Electric Fountain, and the halls on Transportation and Mines. On the opening day, President Grover Cleveland was on hand to push a button that turned on an abundance of electric lights and motors. (Electric light bulbs and ac motors were still fairly new.) This caused such an excited forward surging of the thousands of spectators that many fainted in the crush. One may safely assume that few among the twenty-seven million attendees during the Exposition worried about, say, the ill effects of rapid industrialization. And few if any would have guessed that, just a century later, at a World's Fair held in South Korea, the official U.S. exhibit, as if in obeisance to a new *Zeitgeist*, was reportedly dedicated entirely to the detritus of the post-industrial world, featuring mounds of broken machinery and views of festering nuclear disposal sites; or that the new permanent exhibition at the Smithsonian Institution's Museum of American History, "Science in American Life," devotes much of its space to an exposé of the hazards of science and the public's alleged disillusionment with technology.

Another indication of how much the worldview changed during one century is that one of the major events of the Exposition of 1893 was a spectacular World's Parliament of Religions. Personal religion is, and always has been, close to the hearts of most Americans. But it now seems surprising that on this occasion, in a setting glorifying science and industry, hundreds of religious leaders from all parts of the world met to present their views in two hundred sessions during seventeen days. It was a colorful affair, with Hindus, Buddhists, Jains, Jews, Protestants, Catholics, adherents of Shinto and Zoroaster, and so forth, all meeting together in their robes "for a loving conference," in the words of the chairman of the Parliament, J. H. Barrows. The purpose was clear. As it was for the Exposition as a whole, the subtext of that Parliament of Religions was also progress and harmonious unity. Hence the Exposition, Barrows said, could exclude religion no more than it could exclude electricity. Science was invoked as an ally in reaching a higher unity while serving the needs of mankind.

One of the passionate believers that science, religion, and indeed cultural activities are aspects of one grand unification program was one of the organizers of the Parliament of Religions, Paul Carus, a publisher now remembered mainly for having brought the writings of Ernst Mach to readers in the U.S. The title of his presentation[4] was nothing less than "Science, a Religious Revelation." His was a sort of anticlerical post–Christian Deism, much of which would have appealed to some American statesmen-philosophers of an earlier century. Individual dignity, Carus thought, can only be found through the discovery of truth, and that is the business of science. Hence, he announced, "through science, God speaks to us." One did not have to choose between the Virgin and the Dynamo; rather, the laboratory was the true cathedral, and vice versa. As the masthead of his journal *The Open Court* put it, he was "devoted to the science of religion [and] the religion of science...."

Carus typified a popular, science-favoring universalism of that time which today is severely challenged, from the right and from the left. I have chosen Carus because his world picture was a good example of the then prominent movement, *Modern Monism,* based on the belief in a "unitary world conception." It arose essentially as an anti-thematic response against the Cartesian dualism of the material vs. the mental, and against the multiplicity of common sense experience, with its starting point in personal individuality. The movement on behalf of Monism had the enormous ambition, in the words of Carus, "to direct all efforts at reform, and to regenerate our entire spiritual life in all its various fields." This meant of course replacing conventional religion with what Carus called the "Religion of Truth," where Truth is defined as "the description of fact...ascertainable according to the methods of scientific inquiry." In this sense, "science *is* revelation"; and in this way one would overcome the old, unacceptable dualism of scientific truths vs. religious truths.

The head of the small but ambitious international Monistic movement was the great German chemist Wilhelm Ostwald (Nobel Prize, 1909). Whereas most modern scientists are quite aware of the limits even within their research—as Max Planck said in 1931, "a science is never in a position completely and exhaustively to solve the problem it has to face"— the publications of the monistic movement show that it hoped every aspect of culture, life and society would be guided by monistic ideas, from the education of children to the economy of nations, and of course within the research program of science itself. Thus Ernst Haeckel, another patron of the movement, predicted that physical science would eventually trace back all matter to a "single original element."

Despite the philosophical naiveté of the leaders, the movement attracted for a time an enthusiastic following. In Germany, it had branches in forty-one cities, and even organized public mass demonstrations against the Church. One must perhaps allow for the effects on them of having had to live under the reactionary political clericalism of Germany. But I have intentionally chosen this case of "scientism," of excess on the side of a small minority of scientists, as my first example of *the rhetoric of a polarizing over-reaching by many movements, before and since, on either side.* Thus, caught up in this fervor, Ostwald, with hubris unequaled by the few remaining captives of scientism today, was propelled to the heights of over-ambition, with such claims as these in 1911: "We expect from science the highest that mankind can produce and win on this earth,.... Everything that mankind, in terms of its wishes and hopes, its aims and ideals, combines in the concept God, is fulfilled by science." And finally, "Science, now and with immeasurable success takes the place of the divine." Ostwald added the prophecy that "we see arrive the Monistic Century.... It will inaugurate a new epoch for humanity, just as 2,000 years ago the preaching of the general love for humanity had inaugurated an epoch."[5]

Only a year after this publication, neither the Monistic nor the Christian base for kindness and love of fellow man had triumphed. Instead, war, which William James called the inevitable "bloody nurse of history," had taken charge. Strangely enough, it was Henry Adams who had sensed that the trend would be ultimately against a Monistic Century. Writing in 1905 in his autobiography, *The Education of Henry Adams*, he identified the course of history as away from Unity and toward fragmentation and multiplicity. Indeed, in the aftermath of World War I, the idea of progress, and optimism about the place of science in culture were war casualties. The balance had swung the other way. The only major movement with large political ambition that continued to claim a scientific basis was of course Marxism, especially as defended by Lenin in his 1908 book, *Materialism and Empirio-Criticism*. The assertion that Marxism-Leninism, the founding ideology of the Soviet Union, had anything to do with real science is a rhetorical device, one of this century's great delusions even if this propaganda was taught to every child in Communist countries. It is disproved, not least by the faulty analysis of science and its philosophy in Lenin's own book, and by the widespread mistreatment which Soviet scientists experienced when their theories did not please their government.

Spengler's Prediction of the End of Science

Perhaps the most widely read attack against the claims of optimistic science appeared as the war was ending in 1918, and later it deeply influenced such theoreticians of history as Arnold Toynbee and Lewis Mumford. The book was called *The Decline of the West*, written by a German mathematics teacher, Oswald Spengler. No quick summary can do justice to that richly baroque work, but the point I want to focus on here is what it had to say about the topic before us. Spengler's key conception was that for every part of mankind, in every epoch since Egypt, Greece, and Rome, the history of a civilization has taken fundamentally the same course, and this will continue in the future. Thus our own inevitable destiny in the West is to go to dust according to a timetable that can be calculated from the available precedents. Spengler predicted the very date of our undoubted demise: the year 2000.

The end stages of every civilization, he wrote, can be recognized by the ideas science treasures in its own progress—by the adoption of the notion of causality instead of destiny; by attention to abstractions such as infinite space and to cause and effect, rather than to "living nature." The primacy of the soul is replaced by intellect; mathematics pervades more and more activities; and nature is reinterpreted as a network of laws within the corpus of what Spengler calls "scientific irreligion." Here Spengler introduces his most startling idea, one that has become familiar in new garb also. He warns that it is characteristic of the winter phase of civilization that precisely when high science is most fruitful within its own sphere, the seeds of its own undoing begin to sprout. This is so for two reasons: the authority of science fails both within and beyond its disciplinary limits; and an antithetical, self-destructive element arises inside the body of science itself that eventually will devour it.

The failure of science's authority outside its laboratories, he says, is due in good part to the tendency to overreach and misapply to the cosmos of history the thinking techniques that are appropriate only to the cosmos of nature. Spengler holds that the thought style of scientific analysis, namely "reason and cognition," fails in areas where one really needs the "habits of intuitive perception" of the sort he identifies with the Apollonian soul and the philosophy of Goethe. By asserting that an unbridgeable contrast exists between a pure "rationality" of abstract science and the intuitive life as lived, Spengler commits the same error as all such critics before him and after, to this day, of whom few seem even to have come closer to science than through their school textbooks. Therefore they are ignorant of the vast difference between, on the one hand, "public science"—the final results of intersubjective negotiations to fashion at least a temporary consensus and globalization on the basis of experiment and logic, and on the other hand, the earlier, "private" stage of work in science, where the particular researcher's intuitive, aesthetic, thematic or other non-logical preference may be the key to the individual's advance beyond the previous level of public science. The complementarity between these two quite different stages in the actual development of any scientific result explains why in any given field the findings by natural scientists, operating within vastly different cultures and styles, are eventually harnessed into common products with global validity.

All this may be clear enough to practicing scientists. But, Spengler continues, even in the cosmos of nature there is an attack on the authority of science, arising from within its own empire: Every conception is at bottom "anthropomorphic," and each culture incorporates this burden in the key conceptions and tests of its own science, which thereby become culturally conditioned illusions. All our rushing after positive scientific achievements in our century only hides the fact, he thinks, that as in classical times, science is once more destined to "fall on its own sword," and so will make way for a "second religiousness."

What Spengler termed the orgy of two centuries of exact sciences would shortly be ending, together with the rest of what was valuable in Western civilization. As a kind of postscript, Spengler added his opinion in his later book, *Man and Technics* (1931), that advancing technology, with its mindlessly proliferating products, will also turn out to undermine the society of the West—because, he prophesied, its interest in and support of science and engineering will decrease: the "metaphysically exhausted" West will not maintain advances in these fields. Instead, the previously overexploited races in the rest of the world, "having caught up with their instructors," will surpass them and "forge a weapon against the heart of the Faustian [Western] Civilization." The non-Caucasian nations will adopt the technical skills, excel in them, and turn them against the Caucasian originators. In short, as H. Stuart Hughes put it, Spengler's prediction was that the East will triumph through better technology, first in commerce, and then militarily.[6]

A "Scientific World Conception"—the Vienna Circle

The early response to Spengler's diagnosis was predictably bimodal—on one side there was wide and enthusiastic acceptance, which continues among people today who have never read Spengler but, so to speak, have imbibed his ideas with their mother's milk. On the other side, the opponents of Spenglerian scenarios included of course many prominent scientists. Some of these had joined in a study group that called itself the Vienna Circle, which met in the 1920s and early '30s for discussion and publication. It included Moritz Schlick, Rudolf Carnap, Philipp Frank, Kurt Gödel and Otto Neurath. Among their active sympathizers, they could count Hans Reichenbach and Richard von Mises in Germany, and in America B. F. Skinner, P. W. Bridgman, Charles Morris, and W. V. Quine.

The most influential publication of the core group was a slim pamphlet issued in October 1929 as a kind of manifesto of the movement, the main title being nothing less than *The Scientific Conception of the World*.[7] The very title was a trumpet blast in the fight to change the balance again, to put science back at the center of modern culture, and against what the booklet called, in the first sentence, the chief alternative, the tendency toward metaphysical and theologizing thought, those old helpmates of the Romantic movement.

Although most of the scholars involved in the Vienna Circle concerned themselves chiefly with the study of the epistemological and logical problems at the foundations of science, there was a clear undercurrent of wider cultural, social, political and pedagogic ambitions as well. For, as the manifesto said, "The attention toward questions of life is more closely related to the scientific world conception than it might at first glance appear....For instance, endeavors toward the unification of mankind, toward a reform of school and education, all show an inner link with the scientific world conception....We have to fashion intellectual tools for everyday life....The vitality that shows itself in the efforts for a rational transformation of the social and economic order permeates the movement for a scientific world conception, too" (pp. 304–305).

The members of the Circle associated themselves explicitly not with the Platonists and Pythagoreans, but with the Sophists and Epicureans, "with those who stand for earthly being, and the here and now." A science free from metaphysics would be a unified science; it would know no unsolvable riddles; it would train thinking to produce clear demarcations between meaningless and meaningful discourse, between intellect and emotion, between the areas of scientific scholarship on the one hand and myth on the other. Just as this approach would, by this formulation, clarify the foundations of mathematics, of the physical sciences, of biology and psychology, it would also demystify the foundations of the social sciences, "and in the first place...history and economics." The empiricist, antimetaphysical attitude would help to reject such dangerous conceptions as "folk spirit," and would "liberate one from inhibiting prejudices."

Thus, the "debris of millennia" would be removed, and "a unified picture of this world" would emerge, free from magical beliefs. The social and economic struggles of the time would be ameliorated because the "broad masses of people" would reject the doctrines that have misled them (pp. 315–317). Beyond that, the spirit of the scientific world conception would penetrate "in growing measure the forms of personal and public life, in education, upbringing, architecture, and the shaping of economic and social life according to rational principles." And the manifesto for a new modernity ended with the blazing formulation, in italics: "The scientific world conception serves life, and life receives it" (p. 318).

Perhaps the most carefully developed of the many publications expressing the Circle's position on science and its rationality as the keys to a sane world picture was the major book by Richard von Mises, the Austrian scientist, mathematician, engineer and philosopher (as well as scholar of the poet Rainer Maria Rilke). Von Mises entitled his big volume, with a bit of irony, *Kleines Lehrbuch des Positivismus*:[8] The aim was not only to show what an empiricist-rational scientific world conception would consist of, what its tools would be, and what problems it could solve within the sciences, from mathematics and physics to biology and the social sciences. All this is done in great detail; but an equally motivating force was to present thereby a choice from the then-reigning alternatives in German-speaking Europe: the Kantianism in Germany and the clerical-metaphysical trend in Austria, both of which were then being interspersed with the growing totalitarian ideologies. Von Mises noted his quite explicit opposition to what he called "negativism," in which he includes systematic, philosophical and political anti-intellectualisms that have remained part of the present landscape. Among the examples he cited were, in fact, Oswald Spengler, and the once-popular German philosopher Ludwig Klages, whose point of view was enshrined even in the title of his main work, *The Mind as Enemy of the Soul*.

As a sign that von Mises' main aim of the book was to put science at the center of a healthy culture in the largest meaning of the term, his volume dealt at length with the way the scientific world conception would illuminate the understanding of metaphysics, poetry, art, the law, and ethics. The underlying commonality of the various forms of cultural achievements was considered by von Mises to be due to the principal unity of their methods if carried through rationally and soundly. The original readers of the book must have felt themselves to be in the presence of an updated follower of Auguste Comte. The very last sentence is, as it were, the summary of the whole project: "We expect from the future that to an ever-increasing extent scientific knowledge, i.e., knowledge formulated in a connectable manner, will regulate life and the conduct of man" (p. 370).[9]

Freud: Instinctual Passions versus Reasonable Interests

But now we will see the lever of sentiments shift the balance once more, and indeed on the very issue of whether knowledge formulated in a scientific manner can lead mankind to saner and more rational conduct. In 1929, the same year in which the optimistic manifesto of the Vienna Circle was published, Sigmund Freud, writing in the same city, produced a book of his mature years giving his somber and pessimistic answer. To the founder of psychoanalysis, the role of science in our culture had been a continuing preoccupation, and in 1911 he had still been optimistic enough to sign the *Aufruf* of the Society for Positivistic Philosophy. But in that book of late 1929, *Das Unbehagen in der Kultur*,[10] Freud found that science, while counting among the most visible manifestations of civilization, was at best an ameliorating influence in a titanic struggle on which the fate of our culture depended. That struggle, he said, was centered on mankind's often doomed effort to master "the human instinct of aggression and self-destruction." Even at that time he saw, in the last paragraph of the book, that "Mankind has gained control over the forces of nature to such an extent that with their help it may have no difficulty to exterminate one another to the last man" (p. 92).

Freud held that the restrictions which civilization imposes upon the demands of our instincts produce an irremediable antagonism between those fetters versus the innate "Destructive Instinct," or "Death Instinct" (pp. 7, 8), the drive which is constantly at odds with the civilizing project to elevate the moral condition of mankind. He wrote, "...man's natural aggressive instinct, the hostility of each against all, and of all against each, opposes this program of civilization. This aggressive instinct is the derivative and the main representative of the death instinct which we have found alongside of Eros and which shares world-domination with it. And now, I think, the meaning of the evolution of civilization is no longer obscure to us. It must present the struggle between Eros and Death, between the instinct of life (*Lebenstrieb*) and the instinct of destruction (*Destruktionstrieb*), as it works itself out in the human species. This struggle is what all life essentially consists of, and the evolution of civilization may therefore be simply described as the struggle for life of the human species. And it is this battle of the giants that our nursemaids try to appease with their lullaby about Heaven" (p. 69).

In this conflict, scientific and other cultural activities result from the successful if incomplete "sublimation of instinctual aims," making science at first glance merely a "vicissitude which has been forced upon the instincts by civilization." The accomplishments of science and technology originated as welcome tools in the effort to protect men against the hostile forces of nature; they have now become "cultural acquisitions" that "do not only sound like a fairy tale, they are actual fulfillments of every—or almost every—fairy tale wish." They verge on our attaining the old ideas of "omnipotence and omniscience." Man "has, as it were, become a kind of prosthetic God" (pp. 38–39).

But there's the rub; gladness still eludes him. "Present-day man does not feel happy in his God-like character," either individually or in terms of the group. That again has its reason in the fact that "civilization is built upon a renunciation of instinct," such as sexuality and aggressiveness, and "presupposes precisely the nonsatisfaction (by suppression, repression, or some other means) of powerful instincts." Hence, the "cultural frustration" (*Unbehagen*) which dominates the whole field of social relationships between human beings (pp. 43–44, 62).

Freud's pessimistic conclusion follows: "In consequence of this primary mutual hostility of human beings, civilized society is perpetually threatened with disintegration. The interest of work in common would not hold it together; instinctual passions are stronger than reasonable interests....In spite of every effort these endeavors of civilization have not so far achieved very much....It is always possible to bind together a considerable number of people in amity, so long as there are other people left over to receive the manifestations of their aggressiveness," as in religious or ethnic persecution (pp. 59, 61).

During the decades since this was written, modern history has all too often seemed to be the experimental verification of Freud's dark assessments, according to which neither science nor any other cultural activity can fully displace our animal nature from its central position, but can only delay the ultimate fate that threatens.

Scientists as "Betrayers of the Truth"

Let us now turn to the most recent period. We are familiar enough with the fluctuations, during the 1960s and 1970s, of opinion in academe and among the public regarding the interactions of science and society. But starting in the early 1980s, a new and powerful element entered into this discussion which has now been assuming ever greater attention and institutionalization, at least in the United States. The new element, the new force adding to the derogation of the credibility of science, is the insistence from some quarters—which increasingly has fallen on receptive ears among the population—that to a previously completely unrealized degree the pursuit of science is, and has been all along, ever since the days of Hipparchus and Ptolemy, thoroughly corrupt and crooked. Consequently severe measures must be applied to the practice of science from outside. This assertion, which has become louder and louder over the past few years in books, official reports, and hundreds of articles, has spawned dramatic public hearings, the formation of specific government agencies, university bureaucracies, and quite a few careers. The safeguarding of ethical practices and uses of science, of which there has been a long tradition within the scientific community, is now to be put into better and wiser hands.

A striking, pace-setting example of this assertion was the 1982 book by two influential *New York Times* science editors, William Broad and Nicholas Wade. It states its intention in the title on the jacket, *Betrayers of the Truth: Fraud and Deceit in the Halls of Science*,[11] and follows up with the unqualified canon shot of the opening sentence: "This is a book about how science really works." Going far beyond the need to expose the relatively few rotten apples in any barrel, which the scientific community itself has long recognized as necessary if only for the sake of its own health, this kind of rhetoric has become commonplace. As this book and its many followers proclaim, the relatively few, sad cases of real or alleged misbehavior are the litmus test for the whole enterprise; fraud and deceit are depicted as being part of the very structure of scientific research.

Similarly, the report to Congress by the Congressional Research Service, entitled "Scientific Misconduct in Academia," stated that, more and more, "the absence of empirical evidence which clearly indicates that misconduct in science is not a problem...suggests that significant misconduct remains a possibility." Among all the targets to preoccupy those who are charged with timely attention to misconduct damaging our republic, this formulation singles out the conduct of science as being guilty until proved innocent. Moreover, the tendency has recently been to include in the allegation of *scientific* misconduct not only falsification of data, plagiarism and the like, but also the spectrum of misdeeds more common to flawed mankind generally, and for which sanctions have existed, e.g., "use of university resources for inappropriate purposes, sexual harassment, racial discrimination," etc.[12]

Similarly, the Office of Scientific Integrity Review (OSIR) of the Department of Health and Human Services made part of its proposed definition of "misconduct" in science, apart from fabrication, falsification, and plagiarism, "practices that seriously deviate from those that are commonly accepted in the scientific community" (Federal Code:42 C.F.R. 50.102). The intention here may have been to parallel the way the Supreme Court defined obscenity by reference to the current standards of the local community. However, when it comes to making progress in science, some practices contrary to those common at the time have again and again been the very hallmark of needed innovations—from putting mathematics into physics in the seventeenth century, to the introduction of quanta, which pained even the originator, Max Planck himself, and to the more recent innovation of modern teamwork. The proposed definition of misconduct, with its potential for mischief, was one more example of the gap between the culture of science and the culture outside the lab. One should add that to her credit the director of the National Institutes of Health at the time intervened on that point, objecting that such a community standard "would have implicated even the discoverer of penicillin, who serendipitously found good use for bacteria growing in a contaminated lab dish."[13]

The power of the current generalized allegations against the conduct of science has two components. The first is of course the astonishing claim that basic research scientists in considerable numbers are intentionally false to their own most fundamental avowed mission, namely, to the pursuit of truths; in other words, that not just a few apples are rotten, but that the whole barrel is.

Yet, even in the presence of the occasional scandalous misdeeds by a relatively small number of the world's millions of scientific researchers, the vastly overblown allegation of pervasive and ingrained fraud and deceit in science would not have been taken so seriously that in the U.S. the newspapers, college courses, training courses for scientists and physicians, commissions, Congressional committees, scientific societies, and so on, are now massively and expensively preoccupied with the institutionalization of prevention of misconduct in science. The unrelenting accounts of specific incidents, some outrageous, more of them sensationalized, have left some of the public and legislators feeling that a great plague of dishonesty had invaded all academic laboratories. As the journal *Nature* recently noted shrewdly, the current trend is resulting in "a slow— and Hollywood-assisted—erosion of [the scientists'] public image,...[replacing it] in the public mind by a money-grabbing plagiarizing con-artist."[14] *Time* magazine chimed in with an essay on scientists today, beginning with, "Scientists, it seems, are becoming the new villains of Western society." A raft of best-selling books add up the allegations in diatribes that have the frank aim, in the words of Bryan Appleyard's polemic *Understanding the Present: Science and the Soul of Man*, that science must be "humbled." We are, it appears, standing only on the shoulders of dwarfs.

What is getting lost in this avalanche of excitement, and also in the generally poor, even self-flagellating, responses from some scientific institutions, is some thorough inquiry into the actual rate of serious misconduct among scientists, the kind of empirical research that would yield a reasonable estimate of the likely relative rate of incidents. I have found only some scattered, preliminary steps in this direction, but these suggest that in fact the actual rate of misconduct (rather than suspected, alleged, or "perceived" without hard evidence) is remarkably *low*. Among the available, reasonably quantifiable measures is, for example, the National Library of Medicine finding that for the period of 1977 to 1986, when about 2,780,000 articles were published in the world's biomedical literature, 41 of these had to be withdrawn because fraudulent or falsified data appeared in them—a rate of under two one-thousandths of one percent of scientific publications per decade.[15] Other data support the same point. Thus the Food and Drug Administration, responding to allegations or evidence of misconduct in clinical research with investigational new drugs research, submitted 20 cases of suspected fraud or other criminal violations to the U.S. Attorney General's office. These resulted in 13 convictions of clinical investigators—about one per year, on the average.[16]

Nobody does or should condone even a single case. But even if the actual rate were as much as a hundred times greater than these figures indicate, the intellectually most interesting questions would be, first, why science as a whole progresses so well despite being the work of mere human beings; second, how small the number of alleged misconduct is in this field compared with those in others, ranging from the world of finance, law, industry, journalism, and government at every level—including the expensive penchant of Congress to "earmark" pork-barrel "science" projects, unreviewed by peers, at specific institutions ($651 million in Fiscal Year 1994 alone); and third, why the few cases of highly publicized charges of misconduct in science can so severely undermine the trust and confidence of the public and its representatives in the integrity of research in general.

Science as Myth

The answer to those questions is in good part that there is indeed another, reinforcing reason for the widespread success of the assault on the credibility of scientific research. This second line of attack has been opened up by a loose assemblage made up of a branch of contemporary philosophy of science and other humanists, some of the so-called "strong-program" constructivist portion of sociology, of a small subset of the media, of a small but growing number of governmental officials and political aspirants, and of a vocal segment of literary critics and political commentators associated with the avant-garde of the postmodern movement. This is a potent and eloquent collective of just the sort that in the past has successfully challenged the worldview of their time and place.

The overall message evolving over the past decade or two from that direction is no longer based only on stories of unacceptable behavior among a few scientists. The charge has been generalized and made even more serious: Put in starkest terms, the claim is that the most basic fraud committed by the members of the scientific community is their assertion *that there are any truths to be found at all*. For there really is nothing there even to betray and falsify; and consequently, science is inherently not corrigible, even if all misconduct were eliminated.

From that point of view, the business of science is mainly careerist; for example, building and operating expensive institutions that claim to be looking for objectively ascertainable information about entities like quarks and bosons—which, however, are nothing more than "socially constructed" fictions. Against the naive realism that most scientists still embrace, and the agnosticism of the more sophisticated ones, the new critics counterpoise the radical solution: as one sociologist of science put it recently, "There is no Nature; there is only a communication network [among scientists]." The literature in academe is now full of statements such as "science is a useful myth," or "we must abolish the distinction between science and fiction," or "science is politics by other means."[17]

Scientists have tended to adopt the Baconian view that the acquisition of basic knowledge of causes and interrelations of phenomena—by processes not easily predictable or fully understood—can yield power over those of nature's forces which cause our burdens and ills. But now, the new consortium tells us, the arrow really goes the other way: not from knowledge to power, but from power to knowledge, and to a rather questionable knowledge at that. The attempts to find generally applicable, shareable knowledge about what might be called reality—through the use of both the rational and the intuitive faculties of individual scientists, and through their skeptical but collaborative attempt to achieve some consensus—were not only doomed exercises, but ironically they have led to the disasters that have marked this century. The whole modern era, launched under the flag of progress, has only led to tragedy. The extreme over-optimism of a Herbert Spencer or a Friedrich Engels can never be replaced by a soberer conception. Progress is illusion. The globalizing program of science—to find basic unities and harmony transcending the level of apparent variety and discord—is held to be completely contrary to the postmodern drive that celebrates individual variety and the equality of standing of every conceivable style and utterance, every group and competing interest. Ours is the time to face the end of the search for foundations, the "End of the Modern Era"; we are in a state called the "objectivity crisis"—a fashionable phrase found in the titles of learned conferences and in policy-setting documents to be examined shortly.

Together, these slogans of the newly emerging sentiment indicate that the aim is not merely a call for the improvement of practice or for increased accountability, which is appropriate and being pursued through earnest actions, but at bottom is, for the main branch of the movement of critics, the delegitimation of science as one of the valid intellectual forces, a reshaping of the cultural balance, as we shall see in more detail below. In this respect, there is a big difference here compared with the history of *internal* movements of protest, such as those of the logical positivists within philosophy, the Impressionists or Dadaists within art, the modern composers within music, etc. In all those cases, it was some of the best talent in the field that took up the task of renewal. Not so here—the motivating force is not renewal from within, but radical cultural politics from without.[18]

The Romantic Movement's Challenge

Here we meet a clarifying fact: the contest before us is not new, but draws on historic forces of great strength and durability. Therefore it will be instructive to trace some of the individual steps and stages in this remarkable development of the growing new view, so as to make it easier to extrapolate and to preview the new terrain we may have before us. While I can here only point briefly to a few recent milestones, I shall seek documentation in the recent writings of some of the most distinguished thinkers, rather than, say, through representatives of the Dionysian undercurrent.

Our first informant and guide is Isaiah Berlin, widely regarded as a most sensitive and humane historian of ideas. The collection of his essays, published as the fifth volume of his collected papers,[19] opens with a startling dichotomy. He writes: "There are, in my view, two factors that, above all others, have shaped human history in this century. One is the development of the natural sciences and technology, certainly the greatest success story of our time—to this great, and mounting attention has been paid from all quarters. The other, without doubt, consists of the great ideological storms that have altered the lives of virtually all mankind: the Russian revolution and its aftermath—totalitarian tyrannies of both right and left and the explosion of nationalism, racism and, in places, of religious bigotry, which interestingly enough, not one among the most perceptive social thinkers of the nineteenth century had ever predicted" (p. 1). He adds that if mankind survives, in two or three centuries' time these two phenomena will "be held to be the outstanding characteristics of our century, the most demanding of explanation and analysis."

What might the author intend by so juxtaposing these two "great movements"? One's first temptation may be to see a connection through the fact that during World War II the ingenuity and frantic work of scientists among the Allies, supporting the valor of the Allied soldiers, brought an end to the totalitarian tyranny of that period, which might well have triumphed over the democracies and established itself at least throughout Europe.

But such a response would not be to the point here. What is on Isaiah Berlin's mind is quite different. As we follow his eloquent and subtle analysis, it dawns on the reader that science and tyranny, the two polar opposite movements which he holds to have defined and shaped the history of this century, are somehow intertwined—that the development of the modern natural sciences and technology may, *through the reactions against them*, have unintentionally and indirectly contributed to the rise of those "totalitarian tyrannies."

This stunning connection, to be sure, is never explicitly spelled out by the author. But we can glimpse the implicit argument later in the book, in his chapter significantly entitled "The Apotheosis of the Romantic Will: The Revolt against the Myth of an Ideal World." There, Berlin summarizes the chronology of some basic concepts and categories in the Western world, specifically the changes in "secular values, ideals, goals." What commands his attention is the change away from the belief in the "central core of the intellectual tradition...since Plato," and toward a "deep and radical revolt against the central tradition of Western thought" (p. 208), a revolt which in recent times has been trying to wrench Western consciousness into a new path.

The central core of the old belief system, one that lasted into the twentieth century, rested on three dogmas that the author summarized roughly as follows. The first is that "to all genuine questions there is one true answer, all others being false, and this applies equally to questions of conduct and feeling, to questions of theory and observation, to questions of value no less than to those of fact." The second dogma is that "The true answers to such questions are in principle knowable." And the third: "These true answers cannot clash with one another." They cannot be incommensurate, but "must form a harmonious whole," the wholeness being assured by either the internal logic among or the complete compatibility of the elements (pp. 209–211).

Out of these three ancient dogmas both institutionalized religions and the sciences developed to their present form (although one might add that modern scientists, in their practice, have become aware of the need of proceeding antidogmatically, by conjecture, test, refutation, and assaying probability). In their pure state, these systems are utopian in principle, for they are imbued by the optimistic belief, inherent in and derivable from the dogmas, that "a life formed according to the true answers would constitute the ideal society, the golden age." All utopias, Isaiah Berlin reminds us, are "based upon the discoverability and harmony of objectively true ends, true for all men, at all times and places"—and by implication the same is true for scientific and technical progress, which are aspects of our drive toward what he calls "a total solution: that in the fullness of time, whether by the will of God or by human effort, the reign of irrationality, injustice and misery will end; man will be liberated, and will no longer be the plaything of forces beyond his control [such as] savage nature...." This is the common ground shared by Epicurus and Marx, Bacon and Condorcet, the Communist Manifesto, the modern technocrats, and the "seekers after alternative societies" (pp. 212–123).

But, Isaiah Berlin now explains, this prominent component of the modern world picture is precisely what was rejected in a revolt by a two-centuries-old counter movement that has been termed Romanticism or the Romantic Rebellion. From its start in the German *Sturm und Drang* movement of the end of the eighteenth century, it grew rapidly in Western civilization, vowing to replace the ideals of the optimistic program, based on rationality and objectively true ends, by the "enthronement of the will of individuals or classes, [with] the rejection of reason and order as being prison houses of the spirit."

My own favorite summary of the view of science and its disvalue in nineteenth-century literature is the anti-hero in Ivan Turgenev's gripping novel, *Fathers and Sons*. One of the greatest figures of Russian literature, together with Gogol, Dostoevski and Tolstoy, Turgenev was a poet largely in the tradition of nineteenth-century romanticism, inspired by Goethe, Schiller and Byron, among others. *Fathers and Sons* was published in 1861. Its main figure is Yevgeny Vassilevich Bazarov, a university student of the natural sciences, expecting to get his degree as a physician shortly. Being a scientist who "examines everything from a critical point of view," he confesses himself also to be ideologically and politically a nihilist, the natural consequence of not acknowledging any external authority. All talk of love, or the "mystic relationship between a man and a woman," is to him just "romanticism, humbug, rot, art." It would be better to study the behavior of beetles. Even on his vacation he has brought along a microscope and fusses over it "for hours at a time." Reading Pushkin, he says, is for little boys. He thinks it would be much better to start with Ludwig Büchner's *Force and Matter*, a book published in 1855 and embodying such a flagrantly materialistic view that Büchner was forced to resign from his professorship in Germany. (It is, as it turned out later, the very book Albert Einstein singled out in his *Autobiographical Notes* as one of the two or three that most impressed him as a boy, and caused him to turn to the pursuit of science.)

What matters, Bazarov claims, "is that two and two are four—all the rest is nonsense." When he meets a clever and beautiful woman, he startles his friend by saying that hers would be a beautiful body to examine— on a dissection table. As if in revenge, fate brings him to the bedside of a villager dying of typhus, and he is made to help in the post mortem. But he cuts himself with his scalpel, and soon he is on the verge of delirium, a case of surgical poisoning. As he is dying, he tries to keep hold on his kind of reality by asking himself aloud, "Now, what is 8 minus 10?" In short, he is a caricature recognizable throughout literature—except that the figure of the emotionally dysfunctional scientist, from Dr. Frankenstein to the crew of Dr. Strangelove, causes surgical sepsis not only in each of them, but also in all those around them.

Returning to Isaiah Berlin's account, it is striking that, as he notes, no one predicted that the current form of the worldwide Romantic Rebellion would be what dominates "the last third of the twentieth century." The

Enlightenment's search for generalizability and rational order is depicted by the rebels of our time as leading at best to the pathetic Bazarovs of science, and those must be replaced by the celebration of the individual, by flamboyant anti-rationalism, by "resistance to external force, social or natural." In the words of Johann Gottfried von Herder, the rebel shouts: "I am not here to think but to be, feel, live!" (p. 223). Truth, authority and nobility come from having heroically suffered victimization.

This assertion of the individual will over sharable reason has undermined what Isaiah Berlin had called the three pillars of the main Western tradition. The Romantic Rebellion of course has also given us enduring masterpieces of art, music, and literature. But it originated, as it were, as an antithetical mirror image, created in reaction to the very existence of the earlier Enlightenment-based conception. In the apotheosis of the Romantic Will in our time, it glows forth as the alternative, the "romantic self-assertion, nationalism, the worship of heroes and leaders, and in the end.... Fascism and brutal irrationalism and the oppression of minorities" (p. 225). Moreover, in the absence of "objective rules," the new rules are those that the rebels themselves make: "Ends are not...objective values.... Ends are not discovered at all but made, not found but created."

As a result, "this war upon the objective world, upon the very notion of objectivity," launched by philosophers and also through novels and plays, infected the modern worldview; the "romantics have dealt a fatal blow" to the earlier certainties, and have "permanently shaken the faith in universal, objective truth in matters of conduct" (pp. 236–237)—and, he might have added, in science as well. As any revolt does, this one puts before us seemingly mutually incompatible choices. Just as with quite antithetical cases of excess such as Ostwald's, it is again Either/Or, rather than the needed complementarity of mankind's rational, passionate, and spiritual functions. One is reminded here of the fact that extremes tend to meet each other. Thus the poet William Blake, the epitome of the Romantic Rebellion—who called Bacon's, Newton's and Locke's work satanic—composed in his *The Marriage of Heaven and Hell* (1790) one of the "Proverbs" that reveal the credo of so many of the opposing actors in this story to this day: *"The road of excess leads to the palace of wisdom."*

The Romantic Rebellion Infuses State Policy

Other authors provide verification and elaboration of the implications of Berlin's findings, and especially so of the ominous joining in our century of the extremes of a Romantic Rebellion with irrational political doctrines. This was evident in the "Cultural Revolution" in Mao's China, in the USSR, and in other totalitarian systems. To glance at least at one telling example, the historian Fritz Stern has written about the early phases of growth of Nazism in Germany when there arose in the 1920s, in his words, the "cultural Luddites, who in their resentment of modernity sought to smash the whole machinery of culture." The fury over an essential part of the program of modernity, "the growing power of liberalism and secularism," directed itself naturally also against science itself. Julius Langbehn was one of the most widely read German ideologues in the 1920s, and Stern writes of him, "Hatred of science dominated all of Langbehn's thought.... To Langbehn, science signified positivism, rationalism, empiricism, mechanistic materialism, technology, skepticism, dogmatism, and specialization...."

Long before the Nazis assumed governmental power, some German scientists and other scholars demanded that a new science be created to take the place of the old one which they discredited—a new "Aryan science," based on intuitive concepts rather than those derived from theory; on the ether, the presumed residence of the *"Geist"*; on the refusal to accept formalistic or abstract conceptions, which were reviled as earmarks of "Jewish science"; and on the adoption as far as possible of basic advances "made by Germans."

In a classic study,[20] Alan Beyerchen identified some of the other main pillars of Aryan science. There we find themes uncomfortably similar to those that are again fashionable. A prominent part of Aryan science was, of course, that science, as some would now say, is basically a social construct, so that the racial heritage of the observer "directly affected the perspective of his work." Scientists of undesirable races, therefore, could not qualify; rather, one had to listen only to those who were in harmony with the masses, the *"Volk."* Moreover, this *völkisch* outlook encouraged the use of ideologically screened nonexperts to participate in judgments on technical matters (as in the *Volksgerichte*). The international character of the consensus mechanism for finding agreement was also abhorrent to the Nazi ideologues. Mechanistic materialism, denounced as the foundation of Marxism, was to be purged from science and physics was to be reinterpreted to be connected not with the matter but with the spirit. "The Aryan physics adherents thus ruled out objectivity and internationality in science.... Objectivity in science was merely a slogan invented by professors to protect their interests." Hermann Rauschning, president of the Danzig Senate, quoted Adolf Hitler as follows:

We stand at the end of the Age of Reason.... A new era of the magical explanation of the world is rising, an explanation based on will rather than knowledge. There is no truth, in either the moral or the scientific sense.... Science is a social phenomenon, and like all those, is limited by the usefulness or harm it causes. With the slogan of objective science the professoriat only wanted to free itself from the very necessary supervision by the State.

That which is called the crisis of science is nothing more than the gentlemen are beginning to see on their own how they have gotten onto the wrong track with their objectivity and autonomy. A simple question that precedes every scientific enterprise is: Who is it who wants to know something, who is it who wants to orient himself in the world around him?[21]

One issue was how technology, so useful to the state, could be fitted into the Romantic idea. In recent times, many antimodern movements, including Fundamentalist ones, have embraced technology. But Philipp Lenard, an outstanding physicist but a chief cultural hero of Nazi propaganda, spoke for at least a minority when he said that the tendency of scientific results to prepare the ground for practical advances has led to a dangerous notion, that of man's "mastery" of nature: Such an attitude, he held, only revealed the influence of "spiritually impoverished grand technicians" and their "all-undermining alien spirit." This idea, too, had its roots in the centuries-old history of the rise of Romantic thought. Alan Beyerchen summarizes this section with the observation that "the romantic rejection of mechanistic materialism, rationalism, theory and abstraction, objectivity, and specialization had long been linked with beliefs in an organic universe, with stress on mystery [and] subjectivity...."

Because all these excesses were couched in phrases so reminiscent of currently used ones to delegitimate the intellectual authority of science, it is necessary to keep in mind that there is only a common ancestry of these views, rather than a necessarily causal connection between them. This applies also to the next case, as I turn now to the position embraced by another distinguished contemporary icon among humanists, although an advocate rather than an analyst. His writings on this topic are—like those of Oswald Spengler, or the positivists—of interest here not because they now represent majority positions, which they do not, but because they have the potential for wide resonance at a turning point of sentiments. Also, in this case we shall see that the relation between modern natural science and the rise of totalitarianism, which Isaiah Berlin considered to be only the result of an obscene historic counter-reaction, now receives a much more sinister interpretation: the two become directly, causally linked.

This ominous linkage has been argued repeatedly in writings over the past ten years by the Czech poet, playwright, resistance fighter against

Marxist-Leninist oppression, and statesman—Václav Havel. In the passages to be discussed, we will notice that he subscribes to many of the themes discussed in Isaiah Berlin's analysis; but Havel's key point is that totalitarianism in our time was simply the perverse extreme end result of a trend of ideas embodied in the program of science itself. In this sense, Western science gave birth to Communism; and with the fall of the latter, the former has also been irremediably compromised.

Looking back on this century, other Central Europeans might characterize it as the release of the forces of brutal irrationality and bestiality, a reversion to ruthless autocracies in which the fates of millions were sealed by the whims of Kaiser Wilhelm, Hitler, Stalin and their henchmen—rather than being the offspring of organized skepticism and the search for reasoned consensus, which are at the heart of science. But Havel finds the chief sources of trouble in our century to have been the very opposite, namely, the habit—in his words—of "rational, cognitive thinking," "depersonalized objectivity," and "the cult of objectivity." He advises us to take refuge now in unrepeatable personal experience, in intuition and mystery, and the other mainstays of the Romantic Rebellion. I must let him put his case at some length in his own words; for while he eschews the documentation or balanced account of the scholar, he is instead in fine command of the rhetoric of persuasion, the ease of unspecified assertions and generalizations, and of the chief art of the dramatist, the suspension of disbelief. The result, for many of his readers, is hypnotic acquiescence without questioning the generalities and leaps in the prose. The "end of Communism," he writes in one of his most widely quoted essays,

> has brought an end not just to the 19th and 20th centuries, but to the modern age as a whole.

> The modern era has been dominated by the culminating belief, expressed in different forms, that the world—and Being as such—is a wholly knowable system governed by a finite number of universal laws that man can grasp and rationally direct for his own benefit. This era, beginning in the Renaissance and developing from the Enlightenment to socialism, from positivism to scientism, from the Industrial Revolution to the information revolution, was characterized by rapid advances in rational, cognitive thinking. This, in turn, gave rise to the proud belief that man, as the pinnacle of everything that exists, was capable of objectively describing, explaining and controlling everything that exists, and of possessing the one and only truth about the world. It was an era in which there was a cult of depersonalized objectivity, an era in which objective knowledge was amassed and technologically exploited, an era of systems, institutions, mechanisms and statistical averages. It was an era of freely transferable, existentially un-

grounded information. It was an era of ideologies, doctrines, interpretations of reality, an era in which the goal was to find a universal theory of the world, and thus a universal key to unlock its prosperity.

Communism was the perverse extreme of this trend.... The fall of Communism can be regarded as a sign that modern thought—based on the premise that the world is objectively knowable, and that the knowledge so obtained can be absolutely generalized—has come to a final crisis. This era has created the first global, or planetary, technical civilization, but it has reached the limit of its potential, the point beyond which the abyss begins.

Traditional science, with its usual coolness, can describe the different ways we might destroy ourselves, but it cannot offer truly effective and practicable instructions on how to avert them.[22]

A listener might at this point begin by objecting that these passages are built on immense over-generalizations and illogical jumps, just as flawed as those of the extreme Monists were on the other side; or that at least on factual grounds the self-designation of Communist ideology as "scientific" was indeed a fraud. On this last point, the scholar of the history and philosophy of Soviet science, Loren Graham, made the trenchant observation: "In 1992, the playwright and President of independent Czechoslovakia Václav Havel wrote that the fall of communism marked the end of an era, the demise of thought based on scientific objectivity.... Was the building of the White Sea Canal in the wrong place and by the most primitive methods, at the cost of hundreds of thousands of prisoners' lives, the blossoming of rationality? Was the disregard of the best technical specialists' advice in the construction of Magnitogorsk, the Dnieper dam and the Baikal-Amur Railway a similar victory for objectivity? Was the education of the largest army of engineers the world has ever seen—people who would come to rule the entire Soviet bureaucracy—in such a way that they knew almost nothing of modern economics and politics an achievement of science?.... And even long after the death of Stalin, into the 1980s, what was the Soviet insistence on maintaining inefficient state farms and giant state factories, if not an expression of willful dogmatism that flew in the face of a mountain of empirical data?"[23]

But one may doubt if Havel would be interested in reconsidering his position, for the object of his essay is the conclusion, presenting the "way out of the crisis of objectivism," as Havel labels it. Only a radical change in man's attitude toward the world will serve. Instead of the generalizing and objectifying methods that yield shareable, repeatable, inter- or trans-subjective explanations, we must now turn, he says, to the very opposite, which presumably "science" somehow has totally banished from this world, i.e., to "such forces as a natural, unique and unrepeatable experience of the world, an elementary sense of justice, the ability to see things as others do...courage, compassion, and faith in the importance of particular measures that do not aspire to be a universal key to salvation.... We must see the pluralism of the world.... We must try harder to understand than to explain." Man needs "individual spirituality, firsthand personal insight into things...and above all trust in his own subjectivity as his principal link with the subjectivity of the world...."

Despite Havel's hint, in passing, of a possible blending of the "construction of universal systemic solutions" or "scientific representation and analysis" with the authority of "personal experience," so as to achieve a "new, postmodern face" for politics, Havel's identification of the "End of the Modern Era" is not to be understood merely as a plea for some compromise or coexistence among the rival constructs; that much was announced in an earlier and even sharper version of his essay, one which dealt with the place of modern science quite unambiguously and hence deserves careful reading:

[Ours is] an epoch which denies the binding importance of personal experience—including the experience of mystery and of the absolute—and displaces the personally experienced absolute as the measure of the world with a new, man-made absolute, devoid of mystery, free of the 'whims' of subjectivity and, as such, impersonal and inhuman. It is the absolute of so-called objectivity: the objective, rational cognition of the scientific model of the world.

Modern science, constructing its universally valid image of the world, thus crashes through the bounds of the natural world which it can understand only as a prison of prejudices from which we must break out into the light of objectively verified truth.... With that, of course, it abolishes as mere fiction even the innermost foundation of our natural world. It kills God and takes his place on the vacant throne, so that henceforth it would be science which would hold the order of being in its hand as its sole legitimate guardian and be the sole legitimate arbiter of all relevant truth. For after all, it is only science that rises above all individual subjective truths and replaces them with a superior, trans-subjective, trans-personal truth which is truly objective and universal.

Modern rationalism and modern science, through the work of man that, as all human works, developed within our natural world, now systematically leave it behind, deny it, degrade and defame it—and, of course, at the same time colonize it.[24]

Here we see the giant step which Havel has taken beyond Berlin's analysis: It is modern science itself that has been the fatal agent of the modern era; as if to answer Ostwald's excesses, it is held responsible even for deicide.

Many have been moved by Havel's powerful mixture of poetical feeling, theatrical flourish, and the bold waving of an ancient, bloodstained shirt. The summary of his ideas, published conspicuously under the title "The End of the Modern Era,"[25] made an immediate and uncritical impression on readers of the most varied backgrounds. Among them was one person especially well placed to ponder the values of science, and to draw conclusions of great import for the life of science in the United States. Here we arrive at the last and most recent of the stages on the road to the current understanding of the place of science in our culture.

The person so deeply affected by Havel's piece was none other than the distinguished chairman of the U.S. Congress Committee on Science, Space and Technology, and one of the staunchest and most effective advocates of science during his long tenure in the House of Representatives: George E. Brown, Jr. of California. He acknowledged that he had received "inspiration" from Havel's essay, "The End of the Modern Era," and decided to reconsider his role as a public advocate of science. He therefore first wrote a long and introspective essay[26] under the title "The Objectivity Crisis," and then presented it to a group of social scientists in a public session at the annual meeting of the American Association for the Advancement of Science, under the title "The Objectivity Crisis: Rethinking the Role of Science in Society."[27]

Persuaded by Havel's version of the Romantic Revolt, Brown cast about earnestly for the consequences it should have for the pursuit of science in this country. As a pragmatic political leader, he was primarily concerned with how scientific activity may hold on to some legitimacy—by service to the nation in terms of visible "sustainable advances in the quality of life," "the desire to achieve justice" (which he says "is considered outside the realm of scientific considerations"), and all the other "real, subjective problems that face mankind." He now saw little evidence that "objective scientific knowledge leads to subjective benefits for humanity." The privileging of the claim of unfettered basic research is void too, he said, because all research choices are "contextual" and subject to the "momentum of history."

Moreover, science has usurped primacy "over other types of cognition and experience." Here Brown quoted Havel's definition of the "crisis of objectivity" being the result of the alleged subjugation of our subjective humanity, our "sense of justice,...archetypal wisdom, good taste, courage, compassion, and faith," the processes of science "not only cannot help us distinguish between good and bad, but strongly assert that its results are, and should be, value free." In sum, Brown held, it would be all too easy to support more research when the proper solution is instead "to change ourselves." Indeed, he came to the conclusion that "the promise of science may be at the root of our problems." To be sure, the energies of scientists might still find use if they were properly directed, chiefly into the field of education or into work toward "specific goals that define an overall context for research," such as population control. Embracing a form of Baconianism, Brown thus rejected Vannevar Bush's more general vision for science, a rejection I quoted near the beginning of this chapter (see note 2.) Like Havel's, his answer to the question whether science can share a place at the center of modern culture was clearly *No*.

When George Brown presented his ideas to an audience of scientists at the session he had organized and for which he had selected a panel of social scientists,[28] only one of the panel allowed himself to disagree openly, while another of the panelists urged Brown to go even further still: Perhaps not realizing how close he was coming to the *"völkische"* solution tried earlier elsewhere, including in Mao's Cultural Revolution, he seriously suggested that to screen proposals for scientific research funding the federal government form a variation of the National Science Foundation's Board whose membership should contain such nonexperts as "a homeless person [and] a member of an urban gang." No one there dared to raise an audible objection. One felt as if one glimpsed the shape of a possible future. But it is important to note that in early 1994, Mr. Brown, apparently moved by the intellectual objections, such as those given above, voiced to him by one or two scientists, distanced himself from Havel's position. Indeed, no one can fail to agree with him that in the post-Cold-War context, it is "a moral imperative to enlist science and technology in a campaign for a more productive and humane society in which all Americans can enjoy the benefits of an improved quality of life."[29]

In this brief overview, ranging from the trembling pillars of the Platonic tradition of the West to today's so-called "End of the Modern Era" and the "End of Progress," we have identified some of the chief historic trends that have risen and fallen and risen again in the mixture from which the predominant view of an epoch emerges. Today's version of the Romantic Rebellion, while strong in other fields, represents still only a seductive minority view among analysts and science policy makers, coming not up from the grass roots but down from the treetops. However, while it is held among prominent persons who can indeed influence the direction of a cultural shift, the scientists at large, and especially the scientific establishment, have chosen to respond so far merely with quiet acquiescence. If those trends should continue, and the self-designated postmodernists rise to controlling force, the new sensibility in the era to come will be very different indeed from the recently dominant one.

Experts in science policy are now debating what they call the on-going renegotiation of the "social contract" between science and society.[30] One can argue that such a change has been overdue for many reasons, one being that the relatively protected position given to science for a good part of the last five decades has less to do with society's commitment than with the Cold War and with the implicit over-promises regarding spin-offs which, as Don K. Price warned long ago,[31] would eventually come back to haunt scientists. Adding concerns about the state of the economy, and competitiveness, the lack of general scientific literacy, etc., there is much in such a list to help explain the public's readiness for a reappraisal. But by my analysis, such factors act only as catalysts or facilitators of the tidal change that historically are always potentially present in our culture.

Of course, it may turn out that the present version of the Romantic Rebellion will peter out—although I doubt it will. Or it may gain strength, as it did in the nineteenth-century and again at various times in the twentieth, especially when the scientific community itself paid little attention to the course of events. Or at best a new accommodation might gradually emerge, a "third way," based on a concept analogous to complementarity (and also analogous to the complementarity of personal and public science within the practice of research itself). That is, it may at last be more widely recognized, by intellectuals and the masses alike, that the scientific and humanistic aspects of our culture do not have to be opposing worldviews that must compete for exclusive dominance, but are in fact complementary aspects of our humanity that can and do coexist productively (as Samuel Taylor Coleridge put it memorably in chapter 14 of his *Biographia Literaria*: "in the balance or reconciliation of opposite or discordant qualities"). At any rate, historians will watch the next stages of the old struggle to define the place of science in our culture with undiminished fascination—although perhaps also with an uneasy recollection of Oswald Spengler's prophecy for our *fin-de-siècle*, of Sigmund Freud's pessimism, and of Isaiah Berlin's analysis of the trajectory of our modern era.

References

1.   Gerald Holton, *Science and Anti-Science* (Cambridge, MA: Harvard University Press, 1993), chapter 6.
2.   George E. Brown, Jr., quoted in *Science*, vol. 260 (7 May 1993), p. 735.

3. Leszek Kolakowski, *Modernity on Endless Trial* (University of Chicago, 1900), p. 4.

4. John Henry Barrows, ed., *The World's Parliament of Religions*, v. II (Chicago: The Parliament Publication Co., 1893), 978–981.

5. W. Ostwald, *Monism as the Goal of Civilization* (Hamburg: International Committee of Monism, 1913), 37.
   The section that follows is an abstract of much of Chapter 5, "The Controversy over the End of Science," in Ref. 1.

6. H. Stuart Hughes, *Oswald Spengler: A Critical Estimate* (New York: Charles Scribner's Sons, 1952).

7. Rudolf Carnap, Hans Hahn, and Otto Neurath, *Wissenschaftliche Weltauffassung: Der Wiener Kreis* (Vienna: Artur Wolf Verlag, 1929); for an English translation see Otto Neurath, *Empiricism and Sociology* (Dordrecht: Reidel, 1973). The page references are to the English translation; I have made occasional corrections in the translation, as necessary.

8. He allowed a simpler title, *A Study in Human Understanding*, for the English translation (Harvard University Press, 1951).

9. The word "control," used in the English edition, has been corrected to "regulate," which corresponds to the German edition.

10. Seriously mistranslated into English as *Civilization and its Discontents* (New York: W.W. Norton, 1961).

11. New York: Simon & Schuster, 1982.

12. *Science*, v. 26, 1993, p. 1203.

13. As reported in the *Washington Post*, March 20, 1992.

14. *Nature*, v. 367, 6 January 1994, p. 6. Unlike most scientific journals in the U.S., *Nature* has been alert to the likely damage of the imbalance in reporting. See for example John Maddox's editorial of 17 March 1994 in *Nature*, v. 368, p. 185. It is noteworthy that another among the few who have spoken out against the growing tide of easy condemnation is also a trained science journalist, Barbara J. Culliton, in her essay, "The Wrong Way to Handle Fraud in Science," *Cosmos 1994*, pp. 34–35.
   For an argument on the costs to science that may result from the excesses of distrust in science, see Steven Shapin, "Truth, Honesty, and Authority of Science," in the National Academy of Sciences report *Society's Choices: Social and Ethical Decision Making in Biomedicine* (Washington, DC: National Academy Press, 1994).

15. The data were kindly furnished to me by Donald A. B. Lindberg, Director, National Library of Medicine. These cases are quite different from the laudable practice of scientists publishing correction notices when they find it necessary to draw attention to their own unintended errors. Eugene Garfield, "How to Avoid Spreading Error," *The Scientist* 1:9, 1987, reports that "of the 10 million journal items indexed in the *SCI* [Science Citation Index] since its inception, over 50,000 were coded as explicit corrections.... These vary from corrections of simple typographical errors to retractions of and outright apologies for 'bad' data or data that cannot be verified." This indicates a rate of 0.5 percent for such volunteered corrections of errors.

16. In addition, the Office of Research Integrity of the U.S. Public Health Service recently announced that starting about a year ago and looking back, it has found a total of 14 researchers guilty of some form of scientific misconduct—out of about 55,000 researchers receiving PHS support per year. (Private communication of 20 July 1993 from Lyle W. Bivens, Acting Director, ORI.) The cases involved work that ranged over a considerable period; for example, one of them began in 1977. To get a sense of the low yield of the allegations, and the pedestrian rather than sensational nature of most of the cases, see Office of Research Integrity, *Biennial Report 1991–92*, September 1993, U. S. Dept. of Health and Human Services.
   To glimpse the enormous complexity, cost and labor as well as the fragility of the process of adjudicating allegations of scientific misconduct, see for example the 63-page document, obtainable from the U.S. Department of Health and Human Services, entitled "Departmental Appeals Board. Research Integrity Adjudications Panel. Subject: Dr. Rameshwar K.

Sharma, Docket No. A-93-50, Decision No. 1431, Date: August 6, 1993."

17. For a scholarly and even-handed treatment of the spectrum of the varied interests of sociologists of science, see Harriet Zuckerman, "The Sociology of Science" in Neil J. Smelser, ed., *Handbook of Sociology* (Beverly Hills, CA: Sage Publications, 1988), pp. 511—574.

18. For a thoughtful analysis, see John R. Searle, "Rationalism and Realism, What is at Stake?," *Daedalus*, v. 122, (1993) no. 4, pp. 55—83. A recent book that aims to answer the various types of critics is Paul R. Gross and Norman Levitt, *Higher Superstition: The Academic Left and its Quarrels with Science* (Baltimore, MD: The Johns Hopkins Press, 1994). It is also useful for its extensive bibliography.
   Another stimulating resource is Frank B. Farrell's *Subjectivity, Realism and Postmodernism* (New York: Cambridge University Press, 1994).

19. Isaiah Berlin, *The Crooked Timber of Humanity, Chapters in the History of Ideas* (New York: Random House, 1992).

20. Alan Beyerchen, *Scientists under Hitler: Politics and the Physics Community in the Third Reich* (New Haven, CT: Yale University Press, 1977).

21. Hermann Rauschning, *Gespräche mit Hitler* (New York: Europa Verlag, 1940), p. 210. Mussolini expressed himself similarly.

22. "Politics and the World Itself," *Kettering Review* (Summer 1992), pp.9–11. His essay was also printed on March 1, 1992 in the *New York Times* as Havel's OpEd, entitled "The End of the Modern Era."

23. Loren R. Graham, *The Ghost of the Executed Engineer: Technology and the Fall of the Soviet Union* (Cambridge, MA: Harvard University Press, 1993).

24. Reprinted in Jan Vladislav, ed., *Václav Havel, or Living in the Truth* (London: Faber & Faber, 1987), pp.138–39. The passage was written in 1984.

25. On July 4, 1994 Havel used the occasion to repeat at length much of his previous argument, in the service of explaining the present "state of mind [which] is called postmodernism," and the "crisis" to which science has led mankind. The only new part of his speech (published as an OpEd, July 8, 1994, *New York Times*) is that our "lost integrity" might paradoxically be renewed by "a science that is new, postmodern," such as the "anthropic cosmological principle" and "the Gaia hypothesis."
   This was too much even for Nicholas Wade, who wrote a devastating attack on Havel's essay (N. Wade, "Method and Madness: A Fable for Fleas," *New York Times Magazine*, August 14, 1994, p. 18), ending with: "A view of the world built on the anthropic principle and the Gaia hypothesis would not be post-modern science but rather a throwback to the numerology and astrology from which the era of rationalism has still not fully rescued us.... To subvert rationalism into mysticism would be a cure more pernicious than the disease."
   The seduction of being counted among the postmoderns has apparently attracted even a handful of scientists; the chief example given is their postmodernist interest in "the limits of science." However, the lively discussion of that topic began in the 1870s, led by Emile Dubois-Reymond, and it also preoccupied the logical positivists. For other examples of this old problem, see G. Holton and R. S. Morison (eds.), *Limits of Scientific Inquiry* (New York: W. W. Norton, 1978).

26. Published in September 1992 in the *American Journal of Physics* (vol. 60, no. 9), pp. 779–781.

27. Tape recording of the session (February 12, 1993) obtainable from the American Association for the Advancement of Science. George Brown's own "Opening Remarks" were also distributed as a Press Release by his Washington, DC office.

28. At the February 12, 1993 American Association for the Advancement of Science annual meeting.

29. George E. Brown, "New Ways of Looking at U.S. Science and Technology," *Physics Today*, vol. 47 (1994), p. 32
   In a talk on "The Roles and Responsibilities of Science in Post-modern Culture" (February 20, 1994, at another annual meeting of the American Association for the Advancement of

Science), Mr. Brown remarked: "Let me begin by suggesting that the term 'post-modern culture,' sometimes used to describe the present era, is a rubric that comes from the arts and architecture where it had identifiable meaning. To my mind, if the term post-modern is used as a definitional period for policy, politics, or for economic eras, it leads to confusion; and it will not help us to define a point of departure for our discussion here. I hope today's discourse does not get side-tracked on a tedious dissection of post-modernism. I should note, however, that the editorial that appeared in the *New York Times* two years ago entitled 'The End of the Modern Era' by Czech philosopher and playwright Václav Havel, contained several points to which I agree, and have included in previous talks. Although Havel comes to the terms modernism and post-modernism from his artistic and philosophical orientation, I do not subscribe to those labels, in large part because I do not fully understand his use of them."

Similarly, Mr. Brown is one of the few policy makers who has protested Senator Barbara Mikulski's recent edict that federal funding for basic, "curiosity-driven" research be cut back in favor of supposedly quick-payoff "strategic research."

30. See especially Harvey Brooks, "Research Universities and the Social Contract for Science," in Lewis Branscomb, ed., *Empowering Technology: Implementing a U.S. Strategy* (Cambridge, MA: MIT Press, 1993). Brooks has all along been one of the most prescient and observant authors on the place of science in our culture. See for example his essay, "Can Science Survive in the Modern Age?," *Science,* vol. 174 (1971), pp. 21–30.

31. E.g., in Don K. Price, "Purists and Politicians," *Science,* vol. 163 (Jan. 3, 1969), pp. 25–31.

The Public Image of Science

When future generations look back to our day, they will envy our generation for having lived at a time of brilliant achievement in many fields, and not least in science and technology. We are at the threshold of basic knowledge concerning the origins of life and the universe itself. We are near an understanding of the fundamental constituents of matter, of the process by which the brain works, and of the factors governing behavior. We launched the physical exploration of space and have begun to see how to conquer hunger and disease on a large scale. Scientific thought appears to be applicable to an ever wider range of studies. With current technical ingenuity one could in principle hope to implement many of the utopian dreams of the past.

Hand in hand with the quality of scientific work today goes an astonishing quantity. The worldwide output is vast. And the amount of scientific work being done is increasing at a rapid rate, doubling approximately every ten or twenty years. Every phase of daily and national life is being penetrated by some aspect of this exponentially growing activity.

It is appropriate, therefore, that searching questions are now being asked about the function and place of this lusty giant. Just as a person's vigorously pink complexion may alert the trained eye to a grave disease of the circulatory system, so too may a local celebration of the success and growth of science and technology turn out, on more thorough study, to mask a deep affliction of our culture. And indeed, anyone committed to the view that science should be a basic part of our intellectual tradition will soon find grounds for concern.

Some of the major symptoms of the relatively narrow place science, as properly understood, really occupies in the total picture are quantitative. A nation-wide survey found that nearly 40 percent of the men and women who had attended college in the United States confessed that they had not taken a single course in the physical and biological sciences. Similarly, the mass media pay only negligible attention to their substance; the newspapers have been found to give less than 5 percent of their (nonadvertising) space to factual presentations of science, technology, and medicine; and television has even less. In short, all our voracious consumption of technological devices, all our talk about the threats or beauties of science, and all our money spent on engineering development should not draw attention from the fact that the pursuit of scientific knowledge itself is not a strong component of the operative system of general values.

The Atomization of Loyalties

In the qualitative sense, and particularly among intellectuals, the symptoms are no better. One hears talk of the hope that the forces of science may be tamed and harnessed to the general advance of ideas, that the much deplored gap between scientists and humanists may be bridged. But the truth is that both the hopes and the bridges are illusory. The separation—which I shall examine further—between the work of the scientist on the one hand and that of the intellectual outside science on the other is increasing, and the genuine acceptance of science as a valid part of culture is becoming less rather than more likely.

Moreover, there appears at present no force in our cultural dynamics strong enough to change this trend. This is due mainly to the atrophy of two mechanisms by which the schism was averted in the past. First, the common core of their early education and the wide range of their interests were apt to bring scholars and scientists together at some level where there could be mutual communication on the subjects of their individual competence; and second, the concepts and attitudes of contemporary science were made a part of the general humanistic concerns of the time. In this way a reasonable equilibrium of compatible interpretations was felt to exist, during the last century, between the concepts and problems of science on the one hand and of intelligent common sense on the other; this was also true with respect to the scientific and nonscientific aspects of the training of intellectuals. Specialists, of course, have always complained of being inadequately appreciated; what is more, they are usually right. But although there were some large blind spots and some bitter quarrels, the two sides were not, as they are now in danger of coming to be, separated by a gulf of ignorance and indifference.

It is not my purpose here to urge better science education at the expense of humanistic and social studies. On the contrary, the latter do not fare much better than science does, and the shabby effort devoted to science is merely the symptom of a more extensive sickness of our educational systems. Nor do I want to place all blame on educators and publicists. Too many scientists have forgotten that, especially at a time of rapid expansion of knowledge, they have an extra obligation and opportunity with respect to the wider public, and that some of the foremost researchers took great pains to write expositions of the essence of their discoveries in a form intended to be accessible to the nonscientist. And in the humanities, too many contributors and interpreters seem to scoff at Shelley's contention in his *Defence of Poetry* that one of the artist's tasks is to "absorb the new knowledge of the sciences and assimilate it to human needs, color it with human passions, transform it into the blood and bone of human nature."

It is through the accumulation of such neglects, just as much as through deterioration in the quantity and quality of instruction given our future intellectual leaders, that the acceptance of science as a meaningful component of our culture has come to be questioned. Again, this process is to a large extent merely one aspect of the increasing atomization of loyalties within the intelligentsia. The writer, the scholar, the scientist, the engineer, the teacher, the lawyer, the politician, the physician—each now regards himself or herself first of all as a member of a separate, special group of fellow professionals to which almost all allegiance and energy are given; only very rarely does the professional feel a sense of responsibility toward, or of belonging to, a larger intellectual community. This loss of cohesive links is perhaps the most relevant symptom of the disease of our culture, for it points directly to one of its specific causes. As in other cases of this sort, this is a failure of image.

Pure Thought and Practical Power

Each person's image of science may differ in detail from that of the next, but all public images are in the main based on one or more of seven general positions. The first of these goes back to Plato and portrays science as an activity with double benefits: science as pure thought helps the mind find truth, and science as power provides tools for effective action. In Book Seven of the *Republic,* Socrates tells Glaucon why the young rulers in the ideal State should study mathematics: "This, then, is knowledge of the kind we are seeking, having a double use, military and philosophical; for the man of war must learn the art of number, or he will not know how to array his troops; and the philosopher also, because he has to rise out of the sea of change and lay hold of true being.... This will be the easiest way for the soul to pass from becoming to truth and being."

The main flaw in this image is that it omits a third vital aspect. Science has always had in addition a metaphoric function—that is, it generates an important part of a whole culture's symbolic vocabulary and provides some of the metaphysical bases and philosophical orientations of our ideology. As a consequence the methods of argument of science, its conceptions and its models, have permeated first the intellectual life of the time, then the tenets and usages of everyday life. All philosophies share with science the need to work with concepts such as space, time, quantity, matter, order, law, causality, verification, reality. Our language of ideas, for example, owes a great debt to statistics, hydraulics, and the model of the solar system. These have furnished powerful analogies in many fields of study. Guiding ideas—such as conditions of equilibrium, centrifugal and centripetal forces, conservation laws, feedback, invariance, complementarity—enrich the general storehouse of imaginative tools of thought. Ideas emerging from science are, and will continue to be as they have been since the seventeenth century, a central part of modern culture—through pure thought, through practical power, and through metaphoric influence. A sound image of science must embrace each of the three functions; but usually only one of the three is recognized. For example, folklore often depicts the life of the scientist either as one of pure thought—isolated from practical matters and from its metaphoric meaning—or, at another extreme, as dedicated chiefly to technological improvements.

Iconoclasm

A second image of long standing is that of the scientist as iconoclast. Indeed, almost every major scientific advance has been interpreted—either triumphantly or with apprehension—as a blow against religion. To some extent this position was created by the ancient tendency to prove the existence of God by pointing to problems which science could not solve at the time. Newton thought that the regularities and stability of the solar system proved it "could only proceed from the counsel and dominion of an intelligent and powerful Being," and the same attitude governed thought concerning the earth's formation before the theory of geological evolution, concerning the descent of man before the theory of biological evolution, and concerning the origin of our galaxy before modern cosmology. The advance of knowledge therefore made inevitable an apparent conflict between science and religion. It is now clear how large a price had to be paid for a misunderstanding of both science and religion: to base religious beliefs on an estimate of what science *cannot* do is as foolhardy as it is blasphemous.

The iconoclastic image of science has, however, other components not ascribable to a misconception of its function. For example, the historian Arnold Toynbee charged science and technology with usurping the place of Christianity as the main source of new cultural symbols. Neo-orthodox theologians have called science the "self-estrangement" of man because it carries him with idolatrous zeal along a dimension where no ultimate—that is, religious—concerns prevail. Similarly, T. S. Eliot proclaimed[1] that culture and religion are "different aspects of the same thing" and defined culture and religion in such a way that science, when mentioned at all, would become identifiable with idolatry. It is evident that these views fail to recognize the multitude of divergent influences that shape a culture, or a person. And on the other hand there is, of course, a group of scientists, though not a large one, which really does support the view that science is an iconoclastic activity. Ideologically they are descendants of Lucretius, who wrote on the first pages of *De Rerum Natura,* "The terror and darkness of mind must be dispelled not by the rays of the sun and glittering shafts of day, but by the aspect and the law of nature; whose first principle we shall begin by thus stating, nothing is ever gotten out of nothing by divine power."

Ethical Perversion

The third image of science is that of a force which can invade, possess, pervert, and destroy a person. The current stereotype of the soulless, evil scientist is the psychopathic investigator in science fiction, or the nuclear destroyer—immoral if he develops the weapons the government asks to be produced, almost traitorous if he refuses that request. According to this view, scientific morality is inherently negative. It causes the arts to languish, it blights culture, and when applied to human affairs, it leads to regimentation and to the impoverishment of life. Science is the serpent seducing us into eating the fruits of the tree of knowledge—thereby dooming us.

The fear behind this attitude is not confined to science; it is directed against all thinkers and innovators. Society has always found it hard to deal with creativity, innovation, and new knowledge. And since science assures a particularly rapid, and therefore particularly disturbing, turnover of ideas, it remains a prime target of suspicion. Factors peculiar to our time intensify this suspicion. The discoveries of "pure" science often lend themselves readily to widespread exploitation through technology. The products of technology—whether they are better vaccines, better gadgets or better weapons—have the characteristics of frequently being very effective, easily made in large quantities, easily distributed, and very appealing to the masses. Thus we are in an inescapable dilemma—irresistibly tempted to reach for the fruits of science, yet, deep inside, aware that our metabolism may not be able to cope with this ever-increasing appetite.

The fear that this dilemma can no longer be resolved increases the anxiety and confusion concerning science. A symptom is the popular identification of science with the technology of weapons. Efforts to convince people that science offers us chiefly knowledge about ourselves and our environment, and occasionally a choice of action, have been rather ineffective. On their side, scientists *as scientists* feel they can take little credit or responsibility either for facts they discover—for they did not create them—or for the uses others make of their discoveries, for they generally are neither permitted nor specially fitted to make these decisions. Those are controlled by considerations of ethics, economics, or politics and therefore are shaped by the values and historical circumstances of the whole society. It is, however, also appropriate to say here that there has been only moderate success in persuading the average scientists, engaged in basic research, of the proposition that the privilege of pursuing a field of knowledge with few restraints imposes on them, in their capacity as citizens, a proportionately larger burden of civic responsibility.

There are other evidences of the widespread notion that science itself cannot contribute positively to culture. Toynbee, for example, gave a long list of "creative individuals," from Xenophon to Hindenburg and from Dante to Lenin, in which he did not include a single scientist. In casual conversation among educated people, it is not fashionable to confess to a lack of acquaintance with the latest ephemera in literature or the arts; but one may even exhibit a touch of pride in professing ignorance of science, of the structure of the universe or one's own body, of the behavior of matter or one's own mind.

The Sorcerer's Apprentice

The last two views hold that man is inherently good and science evil. The next image is based on the opposite assumption—that man cannot be trusted with scientific and technical knowledge. He has survived only because he lacked sufficiently destructive weapons; now he can immolate his world. Science, indirectly responsible for this new power, is here considered ethically neutral. But man, like the sorcerer's apprentice, can neither understand this tool nor control it. Unavoidably he will bring catastrophe on himself, partly through his natural sinfulness, and partly through his lust for power, of which the pursuit of knowledge is a manifestation. It was in this mood that Pliny deplored the development of projectiles of iron for purposes of war: "the most criminal artifice that has been devised by the human mind; for, as if to bring death upon man with still greater rapidity, we have given wings to iron and taught it to fly. Let us, therefore, acquit Nature of a charge that belongs to man himself."

When a science is viewed in this plane—as a temptation for the mischievous savage—it becomes easy to suggest a moratorium on science, a period of abstinence during which humanity somehow will develop adequate spiritual or social resources for coping with the possibilities of inhuman uses of modern technical results. Here I need point out only the two main misunderstandings implied in recurrent calls for a moratorium.

First, science of course is not an occupation, such as working on an assembly line, that one may pursue or abandon at will. For a creative scientist, it is in large part not a matter of free choice what he shall do. Indeed it is erroneous to think of him as advancing toward knowledge; it is, rather, knowledge which advances toward him, grasps him, and overwhelms him. Even the most superficial glance at the life and work of a Kepler, a Dalton, or a Pasteur would clarify this point. It would be well if in his or her education each person were shown by example that the driving power of creativity is no less strong and sacred for the scientist than for the artist.

The second point can be put equally briefly. In order to survive and to progress, mankind cannot know too much. Salvation can hardly be thought of as the reward for ignorance. Man has been given his mind in order that he may find out where he is, what he is, and who he is, and how he may assume the responsibility for himself which is the chief obligation incurred in gaining knowledge. For example, in the search for the causes and prevention of aggression among people and nations, we shall find the natural and social sciences to be main sources of understanding.

Ecological Disaster

A change in the average temperature of a pond or in the salinity of an ocean may shift the ecological balance and cause the death of a large number of plants and animals. The fifth prevalent image of science similarly holds that while neither science nor man may be inherently evil, the rise of science happened, as if by accident, to initiate a change in the balance of beliefs and ideas that now corrodes the only conceivable basis for a stable society. In the words of Jacques Maritain, the "deadly disease" science set off in society is "the denial of eternal truth and absolute values."

The main events leading to this state are usually presented as follows: the abandonment of geocentric astronomy implied the abandonment of the conception of the earth as the center of creation and of man as its ultimate purpose. Then the idea of purposive creation gave way to blind evolution. Space, time, and certainty were shown to have no absolute meaning. All *a priori* axioms were discovered to be merely arbitrary conveniences. Modern psychology and anthropology led to cultural relativism. Truth itself has been dissolved into probabilistic and indeterministic statements. Drawing upon analogy with the sciences, other scholars have become increasingly relativistic, denying either the necessity or the possibility of postulating immutable verities, and so have undermined the old foundations of moral and social authority on which a stable society must be built.

Many applications of recent scientific concepts outside science merely reveal an ignorance about science. For example, relativism in nonscientific fields is generally based on farfetched analogies. Relativity theory, of course, does not find that truth depends on the point of view of the observer but, on the contrary, reformulates the laws of physics so that they hold good for all observers, no matter how they move or where they stand. Its central meaning is that the most valued truths in science are independent of the point of view. Ignorance of science may be also the excuse for adopting rapid changes within science as models for antitraditional attitudes outside science. In reality, no field of thought is more conservative than science. Each change necessarily encompasses previous knowledge. Science grows like a tree, ring by ring. Einstein did not prove the work of Newton wrong; he provided a larger setting within which some limitations, contradictions, and asymmetries in the earlier physics disappeared.

But the image of science as an ecological disaster can be subjected to a more severe critique. Regardless of science's part in the corrosion of absolute values, have those values really given us a safe anchor? *A priori* absolutes abound all over the globe in completely contradictory varieties. Most of the horrors of history have been carried out under the banner of some absolutist philosophy, from the Aztec mass sacrifices to the *auto-da-fe* of the Inquisition, from the massacre of the Huguenots to the Nazi gas chambers. Few would argue that any society of past centuries provided a meaningful and dignified life for more than a small fraction of its members. If, therefore, some of the new philosophies, inspired rightly or wrongly by science, point out that absolutes have a habit of changing in time and of contradicting one another, if they invite a re-examination of the bases of social authority and reject them when those bases prove false (as did the colonists in this country), then one must not blame a "relativistic" philosophy for bringing out these faults. They were there all the time.

In the search for a new and sounder basis on which to build a stable world, science will be an indispensable partner. We can hope to match the resources and structure of society to the needs and potentialities of people only if we know more about man. Already science has much to say that is valuable and important about human relationships and problems. By far the largest part of the total research and development effort in science and engineering today is concerned, indirectly or directly, with human needs, relationships, health, and comforts. Because of the large gap between continuing needs and what has been accomplished so far, and also because of the often slanted and unfair distribution of the fruits of advances, giving credit for real achievements in alleviating human misery is often lagging. Here I am thinking, for example of the role of science in the expansion of the idea of human rights, especially by disproving the old pseudo-scientific bases for racism, which were exposed as a sham by the work of anthropologists during this century. One tends to forget that not all the desirable "applications of science" look like VCRs or pills. Insofar as absolutes are to help guide mankind safely on the dangerous journey ahead, they surely should be at least strong enough to stand scrutiny against the background of developing factual knowledge.

Scientism

While the last four images imply a revulsion from science, scientism may be described as an addiction to science. Among the signs of scientism—which afflicts a small part of the general population but also some scientists[2]—are the habit of dividing all thought into two categories, up-to-date scientific knowledge and nonsense; the view that the theoretical sciences and the large laboratory offer the best models for successfully employing the mind or organizing any effort; and the identification of science with technology, to which reference was made above.

One main source for this attitude is evidently the persuasive success of recent technical work. Another resides in the fact that we are passing through a period of fundamental change in the nature of scientific activity—a change triggered by the perfecting and disseminating of the methods of basic research by teams of specialists with widely different training and interests. Until World War II, the typical scientist worked alone or with a few students and colleagues. Today he or she increasingly belongs to a sizable group. Science has increasingly become a large-scale operation with a potential for rapid and worldwide effects. The result is often a splendid advancement in knowledge, but with side effects which are analogous to those of sudden urbanization—a strain on communication, the rise of an administrative bureaucracy, the depersonalization of some human relationships.

To a large degree, this may be unavoidable. The new style of doing science will justify itself by the flow of new knowledge and of material benefits. A danger—and this is the point where scientism enters—is that the fascination with the mechanism of this successful enterprise may change the scientists and society around them. For example, the unorthodox, often withdrawn individual, on whom many great scientific advances have depended in the past, does not fit well into the new system. And society will be increasingly faced with the seductive urging of scientism to adopt generally what is regarded—often erroneously—as the pattern of organization of the new science. The crash programs, the breakthrough pursuit, the megaton effect are becoming ruling ideas in complex fields such as education, where they may not be applicable.

Magic

Few nonscientists would suspect a hoax if it were suddenly announced that a stable chemical element lighter than hydrogen had been synthesized, or that a manned observation platform had been established at the surface of the sun. To most people it appears that science knows no inherent limitations. Thus, the seventh image depicts science as magic, and the scientist as wizard, *deus ex machina,* or oracle. The attitude toward scientists on this plane ranges from terror to sentimental subservience, depending on what motives one ascribes to them.

The Impotence of the Modern Intellectual

The prevalence of these false images is a source of the alienation between the scientific and nonscientific elements in our culture, and therefore the failure of image is important business for all of us. Now to pin a good deal of the blame on the insufficient instruction in science received by the general student at all levels is quite justifiable. I have implied the need for science education, and most people nowadays seem to come to this conclusion anyway. But this is not enough. We must consider the full implications of the fact that not only the man and woman in the street but almost all of our intellectual and political leaders today know at most very little about science. And here we come to the central point underlying the analysis made above: the chilling realization that our intellectuals, for the first time in history, are losing their hold on understanding of the world. The wrong images would be impossible were they not anchored in two kinds of ignorance. One kind is ignorance on the basic level, that of *facts*—what biology says about life, what chemistry and physics say about matter, what astronomy says about the development and structure of our galaxy, and so forth. Nonscientists realize that the old, commonsense foundations of thought about the world of nature have become obsolete. The simple interpretations of solidity, permanence, and reality have been washed away, and they are plunged into the nightmarish ocean of four-dimensional continua, probability amplitudes, indeterminacies, and so forth. They know only two things about the basic conceptions of modern science: that they do not understand them, and that they are now so far separated from them that they may never find out their meaning.

On the second level of ignorance, the contemporary intellectual knows just as little about the way in which the main facts from the different sciences fit together in a world picture taken as a whole. He has had to leave behind him, one by one, those great syntheses which used to represent our intellectual and moral home—the world view of the book of Genesis, of Homer, of Dante, of Milton, of Goethe. All too many find themselves abandoned in a universe which seems a puzzle on either the factual or the philosophical level. Of all the effects of the separation of culture and scientific knowledge, this feeling of bewilderment and basic homelessness is the most terrifying. Here is one reason, it seems to me, for the frequent self-denigration of contemporary intellectuals. Nor are the scientists themselves unaffected, for it has always been, and must always be, the job of the humanist to construct and disseminate the meaningful total picture of the world.[3]

To illustrate this point concretely, and show it is thought by some to apply since the beginning of modern science, we may turn to a widely and properly respected work by E. A. Burtt, a scholar who understood both the science and the philosophy of the sixteenth and seventeenth centuries. The reader is carried along by his authority and enthusiasm. And then, suddenly, one encounters a passage unlike any other in the book, an anguished cry from the heart:

> It was of the greatest consequence for succeeding thought that now the great Newton's authority was squarely behind that view of the cosmos which saw in man a puny, irrelevant spectator (so far as a being, wholly imprisoned in a dark room, can be called such) of the vast mechanical system whose regular motions according to mechanical principles constituted the world of nature. The gloriously romantic universe of Dante and Milton, that set no bounds to the imagination of man as it played over space and time, had now been swept away. Space was identified with the realm of geometry, time with the continuity of number. The world that people had thought themselves living in—a world rich with color and sound, redolent with fragrance, filled with gladness, love and beauty, speaking everywhere of purposive harmony and creative ideals—was crowded now into minute corners in the brains of scattered organic beings. The really important world outside was a world hard, cold, colorless, silent, and dead; a world of quantity, a world of mathematically computable motions in mechanical regularity. The world of qualities as immediately perceived by man became just a curious and quite minor effect of that infinite machine beyond. In Newton, the Cartesian metaphysics, ambiguously interpreted and stripped of its distinctive claim for serious philosophical consideration, finally overthrew Aristotelianism and became the predominant world-view of modern times.[4]

For once, the curtain usually covering the dark fears modern science engenders is pulled away. This view of modern man as a puny, irrelevant spectator lost in a vast mathematical system—how far this is from the exaltation that Kepler found through scientific discovery: "Now man will at last measure the power of his mind on a true scale, and will realize that God, who founded everything in the world on the norm of quantity, also has endowed man with a mind which can comprehend these norms!" Was not the universe of Dante and Milton so powerful and "gloriously romantic" precisely because it incorporated, and thereby rendered meaningful, the contemporary scientific cosmology alongside the moral and aesthetic conceptions? Leaving aside the question of whether most of Dante's and Milton's contemporaries by and large were really living in a rich and fragrant world of gladness, love, and beauty, it is fair to speculate that if our new cosmos is felt to be cold, inglorious, and unromantic, it is not the new cosmology which may be at fault, but the absence of new Dantes and Miltons.

To take another concrete example, consider the widely read book by Arthur Koestler entitled *The Sleepwalkers*. In it, Koestler tried to trace the rise of modern physics (and, with it, of modern philosophical thought), stemming from the work of Kepler, Galileo, Newton, and some of their contemporaries. This is indeed still a useful task to set oneself. Koestler worked with devotion on his material. And, most important, he was of course the intelligent layman *par excellence* whom any scientist would be pleased and proud to have as a colleague and fellow student in an evidently earnest search for an understanding of modern science.

And yet, something terrifying happened as Koestler came to the end of his book. He had been able to see meaning and order in the physics of the seventeenth century. When he turned in the Epilogue to modern physics, however, all sense of understanding and coherence disappeared, and the incomprehensible modern conceptions seemed to rise around him on every side as threats to his sanity. As he summarizes his work, he finds that to a large degree "the story outlined in this book will be recognized as a story of the splitting-off, and subsequent isolated development, of various branches of knowledge and endeavour—sky-geometry, terrestrial physics, Platonic, and scholastic theology—each leading to rigid orthodoxies, one-sided specializations, collective obsessions, whose mutual incompatibility was reflected in the symptoms of double-think and 'controlled schizophrenia.'"[5]

I believe it is important to consider this case as sympathetically as we can—to listen to the anguish of an intelligent man who has discovered that he cannot cope with the modern conceptions of physical reality. For what he is saying is what most people would say—if they were eloquent enough and interested enough in knowledge to be deeply disturbed by a state of threatening ignorance:

Each of the 'ultimate' and 'irreducible' primary qualities of the world of physics proved in its turn to be an illusion. The hard atoms of matter went up in fireworks; the concepts of substance, force, of effects determined by causes, and ultimately the very framework of space and time turned out to be as illusory as the "tastes, odours and colours" which Galileo had treated so contemptuously. Each advance in physical theory, with its rich technological harvest, was bought by a loss in intelligibility....

Compared to the modern physicist's picture of the world, the Ptolemaic universe of epicycles and crystal spheres was a model of sanity. The chair on which I sit seems a hard fact, but I know that I sit on a nearly perfect vacuum.... A room with a few specks of dust floating in the air is overcrowded compared to the emptiness which I call a chair and on which my fundaments rest....

The list of these paradoxa could be continued indefinitely; in fact the new quantum-mechanics consist of nothing but paradoxa, for it has become an accepted truism among physicists that the sub-atomic structure of any object, including the chair I sit on, cannot be fitted into a framework of space and time. Words like 'substance' or 'matter' have become void of meaning, or invested with simultaneous contradictory meanings....

> These waves, then, on which I sit, coming out of nothing, travelling through a non-medium in multidimensional non-space, are the ultimate answer modern physics has to offer to man's question after the nature of reality.[6]

At the very end of the book, in its last, furiously splashing paragraph, I cannot but hear the cry of a drowning man, a cry for help that cannot leave one unconcerned if one believes that science can and must be shown to play a valid, creative part within our culture:

> The muddle of inspiration and delusion, of visionary insight and dogmatic blindness, of millennial obsessions and disciplined double-think, which this narrative has tried to retrace, may serve as a cautionary tale against the *hubris* of science—or rather of the philosophical outlook based on it. The dials on our laboratory panels are turning into another version of the shadows in the cave. Our hypnotic enslavement to the numerical aspects of reality has dulled our perception of non-quantitative moral values; the resultant end-justifies-the-means ethics may be a major factor in our undoing. Conversely, the example of Plato's obsession with perfect spheres, of Aristotle's arrow propelled by the surrounding air, the forty-eight epicycles of Copernicus and his moral cowardice, Tycho's mania of grandeur, Kepler's sunspots, Galileo's confidence tricks, and Descartes' pituitary soul, may have some sobering effect on the worshippers of the new Baal, lording it over the moral vacuum with his electronic brain.[7]

Burtt and Koestler correctly reflect the present dilemma—and I could have chosen equally vivid examples from the more recent avalanche of books with the same animus. What their outbursts tell us, in starkest and simplest form, is this: by having let the intellectuals remain in terrified ignorance of modern science, we have forced them into a position of tragic impotence; they are, as it were, blindfolded in a maze through which they cannot traverse. They are caught between their irrepressible desire really to understand this universe and, on the other hand, their clearly recognized inability to make any sense out of modern science. As I noted in the previous essay, the great literary critic and historian Lionel Trilling testified to this problem frankly in his Jefferson Lecture (1971), entitled "The Mind in the Modern World."

Once this is understood, the consequence also becomes plain. The intellectual of tomorrow will probably have an even more distorted image and fearful response with regard to science, for there is at present no countercyclical mechanism at work. Hardly anything being done or planned now is adequate to deal with the possibility of a cultural psychosis engendered by the separation of science and the rest of culture.

Better education cannot be the only remedy; the uses and abuses of science also need reexamination. But making science again a part of every intelligent person's educational resource is the minimum requirement—not because science is more important than other fields, but because it is an integral part of a sound contemporary world view. A plausible program would include thorough work at every level of education—imaginative new programs and curricula[8]; strengthened standards of achievement; extended college work in science, and expansion of opportunity for adult education. Few people have faced the real magnitude of the problem. Moreover, while some time-lag between new discoveries and their wider dissemination has always existed, the increase in degree of abstraction, and in tempo, of present-day science, coming precisely at a time of inadequate educational effort has begun to change a lag into a discontinuity.

This lapse, it must be repeated, is not the fault of the ordinary citizen who can only take his cue from the intellectuals—the scholars, writers, and teachers who deal professionally in ideas. Every great age has been shaped by intellectuals who would have been horrified by the proposition that cultivated men and women could dispense with a good grasp of the scientific aspect of the contemporary world picture. This tradition is broken; very few are now able to act as informed mediators. To restore science to reciprocal contact with the concerns of most persons—to bring science into an orbit about us instead of letting it escape from our intellectual tradition—that is the challenge which intellectuals have to face today.

References

1. T. S. Eliot, *Notes Towards the Definition of Culture* (London: Faber & Faber, 1948).
2. One is reminded of the diagnosis C. P. Snow offered in his provocative book *Science and Government* (Cambridge: Harvard University Press, 1961) for the reason why some scientists so single-mindedly stuck to a narrow decision or were satisfied with a narrow range of investigations. It was their success in one particular field or with the operation of one particular aparatus. Snow dubbed these persons "gadgeteers."
3. We also sorely need to give our young scientists more broad humanistic studies—and if I have not dwelled on this it is because, in principle, this can be done with existing programs and facilities, for the tools of study in the humanities, unlike the tools in science, are still to a large degree in touch with our ordinary sensibilities.
4. Edwin A. Burtt, *The Metaphysical Foundations of Modern Physical Science* (New York: Harcourt, Brace & Co., 1927), pp. 236–237.
5. Arthur Koestler, *The Sleepwalkers* (London: Hutchinson, 1959), p. 518.
6. Ibid., pp. 529–531.
7. Ibid., pp. 541–542.
8. An attempt in this direction is indicated in the essay, "How Can Science Courses Use the History of Science?"

"Doing One's Damnedest": The Evolution of Trust in Scientific Findings

Introduction

Lest the title of my essay mislead, I must declare at the outset that I shall not be addressing the scrutiny of scientists by politicians or the press, the topic currently so fashionable. I have done my share of writing on public trust in scientists, for the history of science does have something to say about it; one of this field's opportunities and obligations is indeed to trace changes in the intellectual and moral position of the community of scientists within society. Over the past decades, a rapid evolutionary process has been at work, shaping both the internal self-regulation and the external bureaucratization of the scientific profession, just as over the past four centuries the concepts, theories and practice of science themselves have undergone a more familiar evolutionary development.

Indeed, a chief point I shall make is that the scientists' own *methods* for finding trustworthy research results are the result of a historic progression through stages, each with its own pitfalls and limits, and that here too we have no reason to think our ways will be adequate for tomorrow's needs. Our mental and physical tools have to develop constantly, if only because it usually becomes harder to take the next step as a field evolves. There is within science a marked difference from one end of a century to the other in what it takes to trust one's own data or to trust the results of others. The same evolutionary process has external effects—among them how research is trusted by those outside the scientific community. Today's standards have been assembled from a steady stream of historically identifiable intellectual and social inventions; by the same token, we should expect the tools of science in the future to be in many ways quite different from today's. It will be wise to let the historical view prepare the mind for the next phase. To this end, I shall select a few telling episodes that illustrate the process of evolution of trust in scientific research results.

The Aesthetics of Necessity

An appropriate beginning is an example taken from the start of the modern period itself, which is usually associated with the work of Nicolaus Copernicus. Until some years ago, the reigning opinion in the history of science was that the Copernican system was the very model of a new theory arising from a crisis brought about by the accumulation of data that contradicted the old and overly complex theory of Ptolemy and his followers. Copernicus's book on the revolutions of the celestial objects, it was said, used more trustworthy observational data, yielded a better and more parsimonious theory, and so rescued the practitioners of the time by giving them at last a calculational method of greater accuracy for astronomical predictions.

But this interpretation is now seen by scholars in the field to be an ahistorical imposition of today's criteria of good behavior at the lab bench. On the contrary, Copernicus is chiefly an exemplar of the early introduction into science of essentially thematic presuppositions—that is, of deep convictions about nature on which the initial proposal and eventual acceptance of some of the most powerful scientific theories are still based.[1] Even if Copernicus's main goal had been to help the calculators of ephemerides, which was not the case, he gravely disappointed them. They were no better off after the publication of his *De Revolutionibus*, and in fact they continued to use the Ptolemaic system in essentially the form Ptolemy himself had set forth. For example, Copernicus's system of 1543 gave the same large errors for the predicted location of Mars—up to 5 degrees—as did the ephemerides of Regiomontanes in the 1470s.

Like Ptolemy, Copernicus selected just enough data (among them many with worse errors than he realized) to get his orbits, even bending some of them by a few minutes of arc as needed. But one must remember what he really wanted to achieve. This he made quite plain near the very beginning of his great book: "to perceive nothing less than the design of the universe and the fixed symmetry of its parts."[2] What is more, he wanted to do so by sticking to what he called "the first principle of uniform motion" (that is to say, Aristotelian circular motion), instead of employing nonuniformity and nonconstancies as the Ptolemaics allowed. What convinced him to make this the cornerstone of his argument, and eventually persuaded his followers, was that he thereby produced a model of the planetary system in which the relative locations and order of orbits were no longer arbitrary but followed by necessity. In short, Copernicus is a case study of the privileging of an aesthetically based theory—above all the aesthetics of necessity—and of the temporary rejection of "data" that would appear to disprove a favored theory.

That daring approach has remained to this day a genetic trait in science, at least during the individual scientist's private, nascent considerations; and with all its dangers, the glorious thing discovered later was that it often works, and works without bending the data. Indeed, it can turn out that when a thematically and intellectually compelling theory is given a chance, better data, gathered with its aid, will eventually reinforce the theory. That is the meaning behind a remark Einstein made before the test of General Relativity: "Now, I am fully satisfied, and I do not doubt any more the correctness of the whole system, may the observation of the eclipse succeed or not. The sense of the thing is too evident." When a discrepancy of up to 10 percent remained between the first set of measured deviations of light and the corresponding calculations based on his theory, he responded, "For the expert this thing is not particularly important, because the main significance of the theory doesn't lie in the verification of little effects, but rather in the great simplification of the theoretical basis of physics as a whole."[3] And even before more data came in that decreased the discrepancies, other scientists joined Einstein's camp, persuaded, in the words of H. A. Lorentz, that his grand scheme had "the highest degree of aesthetic merit; every lover of the beautiful must wish it to be true."[4] While this kind of procedure in science is of course much more risky for ordinary mortals, what most characterizes giants like Copernicus and Einstein as they struggle with problems too vast to be solved by the standard procedure, i.e., induction from good data, is their intuition where science will go next. Scientific intuition, when it works, is a gift for which Hans Christian Oersted provided us the happy term, "anticipatory consonance with nature."[5]

Yet even the giants of science can't count on their sense of "what's right." When Einstein asked himself what caused him to be so obstinately against a belief in the fundamentality of probabilism in physics, which the success of quantum mechanics forced on most others, he admitted to Max Born—as I have noted—that he could not provide logical arguments for his conviction, but could only call on his "little finger as witness." By now we know with fair certainty that in this instance the "little finger" test failed.

Increasing the Probability of Reason's Claim

But I am getting ahead of my story. On the next stage along the line of the evolution of trust in one's results we find Galileo. Galileo's invention amounted to secularizing science, submerging the qualitative in favor of the quantitative as the earmark of truth, and elevating experimental checks from illustrations of the value of a theory to the test of its *probability*. In a famous passage in the Third Day of the *Two New Sciences* of 1638, Galileo's spokesman, Salviati, goes to such lengths to describe one of these experiments in detail—the accelerated motion of a bronze ball down an inclined plane, including his claim to have "repeated it a full hundred times"—that his supposedly skeptical listener Simplicio is made to confess why he trusts Salviati's account: "I would have liked to be present at these experiments; but feeling confidence in the care with which you perform them, and in the fidelity with which you relate them, I am satisfied and accept them as true and valid." To which Salviati quickly and eagerly responds, "Then we can proceed with our discussion."[6]

This is still uncomfortably close to Aristotle's advice, in his *Rhetorica*, on how to persuade one's listener. Proof provided by the speaker is only part of it. Equally important, persuasion "depends on the personal character of the speaker, and on putting the audience in a right frame of mind." A few decades ago, the science historian Alexandre Koyré, unlike the more pliable Simplicio, found Galileo's account of the detailed care he took in conducting his experiments to be so overblown and unconvincing that he questioned whether Galileo had made any experimental checks at all. Instead, he argued, Galileo had presented only *Gedankenexperimente*, or thought experiments; he was not describing real ones at all.[7]

Fortunately for Galileo's reputation, some pages of his experimental tests have been found and analyzed, and they show that he—who once upon a time had been called the Father of Experimentation—did in fact do experiments. But modern analyses of his lab notes show why Galileo's private test results remained unpublished, and why he could not fully trust them to do more than, in his phrase, to *"increase the probability"* of "what reason tells me."[8]

At that early stage of the evolution of trust, Galileo's private calculations were largely limited to proportions, and his published work was still narrative. Not until three decades later, in Newton's *Principia*, does the parsimonious style and axiomatic presentation, modeled on Euclid's geometry, take over—the sparse style of public science to which we are used today. In Galileo's books there was still no use of algebra. Galileo does not announce his famous law of free fall as we do in elementary physics, $s = 1/2 \, gt^2$. Rather, he makes the mathematically equivalent, but seemingly quite mysterious, statement: "So far as I know, no one has yet pointed out that the distances traversed, during equal intervals of time, by a body falling from rest, stand to one another in the same ratio as the odd numbers beginning with unity."[9]

To have put it this way means that what counted for Galileo most was after all not the limited and perhaps rather silly case of a falling stone or a rolling ball, but the demonstration that terrestrial phenomena, of which these are examples, can be explained by the operation of *integers*—just as the Pythagoreans had dreamt (and as quantum physicists have proved in our century for atomic behavior). Galileo, too, was still engaged in a search for cosmic truths, a tendency which, for better or worse, had to be reined in as science evolved further.

Witnessing and Peer Review

The next steps in the evolution of trust were two. They came quickly, and both had to do with the ontological status of a "fact." While Galileo and Newton—a much better experimenter—could still regard their activities in the laboratory mostly as a private affair, others from the middle of the seventeenth century on struggled with the recognition that the door of the lab had to be unlocked; that Simplicio, so to speak, had to be invited to be present while the experiment was being performed. Fact had to be democratized. The academies and the Royal Society devoted much of their energies to the public elicitation of facts, the demonstrations of new phenomena carried on before an audience of interested fellow amateurs, acting as witnesses.[10]

This practice can be still found a century and a half later, in the work of Oersted, who is most remembered for his publication on 21 July 1820 describing the discovery of the interaction of current-carrying conductors and magnets.[11] His hasty and perfunctory laboratory notes of findings, which changed all of physical science, would not be accepted now from a beginning student in an introductory physics course. Nor do they fit the new rules of explicitness that lawyers are busily writing today for government agencies to issue and for us to obey. And Oersted, a true Romantic, was still frank and personal in his publications; he did not know, as Louis Pasteur later told his students, that a scientist has to make the results "look inevitable."

Few modern researchers are likely to admit, as Oersted gladly did, that he had been completely convinced many years earlier of the existence of the effect he eventually discovered. (On the other hand, Oersted thus could in principle have been able to obey a memo of the sort sent out by our Secretary of the Department of Energy in 1989, asking that he be given "advance notice of all significant scientific discoveries.") Oersted had been persuaded of a connection existing between electricity and magnetism by reading Immanuel Kant, who on metaphysical grounds proposed that all the different forces of nature are only different exemplifications of one fundamental force, a *Grundkraft*.

Oersted is also an example of a scientist still on the early rungs of the evolutionary ladder of trustworthy methods, as is demonstrated by his choice of procedure for what we now would call peer review. There being no other physicists of note available in Copenhagen at the time, the peers he selected to vouch for the truthfulness of his report were brought in more for presumed moral authority than for scientific acumen: he assured his readers that the experiments he reported were conducted in the presence of his friend Esmarck, the King's Minister of Justice, Wleugel, Knight of the Order of Danneborg and president of the Board of Pilots, and several other gentlemen whose word, one could assume, had to be trusted.[12]

Deciding What Is a Fact

In the meantime, between Newton and Oersted, another enormously important problem was being wrestled with: how to determine which of all possible demonstrable events are indications of scientifically usable phenomena; which of them are really connected to the fixed regularities of nature, and which are merely passing phantoms, clouds with ever-changing form never twice the same, and thus reflecting only ephemeral concatenations? We might call it the problem of telling the difference between signal and noise.

The matter can be summarized most simply by contrasting the styles of research of Robert Boyle of the second half of the seventeenth century and of Charles DuFay in the early decades of the eighteenth. Boyle, while still proclaiming the superiority of reason over authority and even over experience, argues forcefully in his "Proemial Essay" for the inclusion in scientific reports of as many readings and as much detail of the experimentation as possible. He sounds quite modern when he favors "information of sense assisted and heightened by instruments" or argues that "artificial and designed experiments are usually more instructive than observations of nature's spontaneous acting." And he tries to respond to the warning of Francis Bacon: "Nothing duly investigated, nothing verified, nothing counted, weighed or measured is to be found in natural history; and what in observation is loose and vague is in information deceptive and treacherous."

But Boyle really doesn't yet know the difference between readings and data. His method of measurement is still quite primitive. For example, in his famous experiments on the compressibility of the air, he quietly assumes that the tube in which the gas column is being compressed is of uniform diameter, that the mercury in it is sufficiently degassed, and also that he can make, with high accuracy, naked-eye readings of its level with reference to a paper scale pasted on the outside of the glass tube.

Among Boyle's contemporaries there were a number of fellow amateurs of science and instrument makers who in fact dedicated themselves to increasing the range and accuracy of basic measurements. They typified the sort of person who, from the sixteenth century to our day, has revelled in the invention of more and more ingenious devices for measuring time, distance, angle, or mass. Their achievements have made it possible to obtain more accurate and useful values both of derived quantities (e.g., force, pressure, electric charge) as well as of the physical constants of nature. That quest is itself a rather heroic chapter in the history of science and technology. It recounts such triumphs as the nearly logarithmic rates of *decrease* in the uncertainties of weighing (3 orders of magnitude, from about 1550 to 1950); in the error, in seconds per day, of time measurement (8 orders of magnitude in three centuries, from about 1650); and in the error of astronomical angular measure (5 orders of magnitude, from 1600 to 1920s). The same story is told in the nearly logarithmic rate of *increase* of the resolving power of microscopes (from 0.9 to 0.2 microns, between 1840 and 1880) and of the energy reached for initiating elementary particle reactions (10 orders of magnitude, from about 200 KeV in 1930 to over 1,000 TeV in the mid-1980s).[13]

Taking that route toward greater reliability of results was not in the forefront of Boyle's thought. But more significant for us here is that he was still wide open to anything that might happen before his eyes and was apt to count everything observable as a fact to be used in his research. In his desire to find the gold of trustworthy detail, he was smothered by slag and mere sand. And that is of course one reason why not he but one of his readers, Richard Townley, discovered among Boyle's data what we now call Boyle's law.[14]

With Charles DuFay, as Lorraine Daston has pointed out, we see a representative of this new stage of development. In his researches on electricity, DuFay alters his experimental conditions constantly, with the aim of isolating the relevant variable. He is interested only in those "facts" that, he says, are "characteristic of a large class of bodies, not of isolated species," and can therefore be organized into some scheme or simple rule. In short, he intuits the modern characteristic of facticity: that fact is consensual, invariant, and universal. And he signals the necessary next step toward the modern base of trust in experimental reports: He takes on a collaborator, not merely to witness but to try to repeat what DuFay has observed. DuFay realizes that experiments are very difficult to do right at an early stage of a science. More generally, he insists that phenomena do not assume the status of fact until other investigators go over the same ground, repeat the observations, and give their consent. This practice was a key invention along the path of increasing trust in research results.

The Emergence of Teams

If we had enough space, we would now look at the next contributions to the development of trust as seen in the work of Oersted's followers, particularly Faraday and Maxwell. But let me go on to one of Maxwell's successors at the Cavendish Laboratory, J. J. Thomson, usually identified as the discoverer of the electron in the 1890s, and to his student and successor, Ernest Rutherford. They exemplify something new in the evolving tale of trust within the lab.

With "J.J.," then still in his mid-thirties, as the new professor, the Cavendish Laboratory officially became a graduate school in 1895. Ernest Rutherford was among the first advanced students to arrive at the lab, soon to be followed by others of great talent, including a few guests from the U.S. What occurred there was another step toward the modern phase of science: the forging of a group identity among the participant scientists.

Even before "J.J." established himself as a major charismatic leader, he had his students assist him in his experimental researches and left it to them to chase down the details once he himself was satisfied he had ob-

tained the right order of magnitude from theory. But in addition to his
natural way of mentoring his assistants and being interested in each student's work, he had another, rather English weapon: regular afternoon
teas for all researchers in his room. That was an occasion to develop the
sense of community among the researchers, who not only shared the same
interests but also depended on one another in their work. By and by, there
emerged what Dong-Won Kim in his doctoral thesis[15] identified as a core research group around the professor, though only rarely was the collaboration
with J.J. so close as to merit, in his view, joint authorship of resulting papers.

The whole operation, though successful, was still a hand-to-mouth affair, financed largely by student fees and scholarships. Only inexpensive
materials, such as glassware blown by the students themselves, were
likely to be used. The sociology of the modern laboratory was beginning
to come into view, but not its economics. However, the combination of
the selection of the advanced students and the socialization of the group
as a whole around the more and more famous central figure was so effective that it gave rise to a kind of moral imperative, a bonding. These individuals would mind no hardship and, I should imagine, would rather have
committed ritual suicide than betray the scientific norms of the group and
of the times. They were not separate, career-oriented passers-through but
"citizens" of a community who shared in a common spirit.

Rutherford built on this system with a vengeance when he started his
own operations at McGill, then at Manchester, and finally at the
Cavendish Laboratory in Cambridge. Lab work was now planned more
carefully, progress was discussed with each student at least weekly, and a
list of research tasks that Rutherford thought could be completed with reasonable success was given to each student annually. Almost as a by-product of his campaign to understand the atomic nucleus, Rutherford's approach was an early example of a style that would soon lead to real team
research. Each student still had to make the most of his or her own equipment, and each had been selected by Rutherford with an eye to being
"first class," willing to work hard as if in a constant race and with only the
most minimal funds. But despite the fact that they all labored on different
projects that were linked together only in Rutherford's mind, he expected
full loyalty to himself and to the laboratory. He was ever-present. Some of
his students later reported they felt he had adopted them. Not for nothing
was he known to his students as "Papa," according to J. C. Crowther.
Trust in each other and in the quality of their data was assured as automatically as it would have been in a healthy family or an isolated tribe of
hungry hunters.

Still, the organization of the laboratory was by no means the modern one of team research, that is, research where a whole laboratory, with many participants, works together on a shared project. Nor was the Cavendish operation, while magnificent in its production of major results, "modern" in terms of understanding the relative value of money for apparatus versus the extra time spent on work when equipment is scarce. I once asked James Chadwick, who had discovered the neutron after a painstakingly long search in 1932 at the Cavendish with the most primitive equipment, to describe the atmosphere there at the time. He wrote back that his hope had been to settle the question whether the neutron existed by building a small accelerator. However,

> no suitable transformer was available and, although Rutherford was mildly interested, there was no money to spend on such a wild scheme. I might mention that the research grant was about £2000 a year, little even in those days for the amount of work which had to be supported. I persisted with the idea for a year or two....I had quite inadequate facilities, and no experience in such matters. I wasted my time—but no money.[16]

Wasting a year or two of a Chadwick for lack of budgeted funds was soon to become unacceptable, and so would a style of research that failed to take full advantage of the combined talents of a group. The earliest exemplar of that new style in physics was the "family" of young researchers that Enrico Fermi educated and assembled around himself with spectacular results from about 1929 to the mid-1930s. Here, again, group loyalty was extraordinary. This being Rome, and Fermi being thought infallible in quantum physics, his students called him "Il Papa." Money was no real problem; with excellent political instinct, Fermi had secured the patronage of the Italian state for his laboratory needs. And with equally good managerial instinct he designed methods for deciding on promising research lines and then pursuing them fiercely, with a division of labor within the group that could be a model even now.

All this is implied even in Fermi's decision to list as authors of his laboratory's publications every one of the various members of that team—at least in physics a hitherto unheard-of public assignment of credit and sharing of responsibility for the results. Consider the paper reporting the startling discovery of the artificial radioactivity produced with beams of slow neutrons, sent to the *Ricerca Scientifica* on 22 October 1934. This paper can be said to have effectively opened the nuclear age by announcing the discovery of resonance and introducing the concept of the moderator for nuclear reactions, and it figured in the award to Fermi of the Nobel prize for physics four years later.[17]

There were five authors, given in this order: Fermi, Amaldi, Pontecorvo, Rasetti, and Segrè. We know from their various biographical statements that the young associates of Fermi worked closely under his inspiration, and the listing of authors, with the name of Fermi put first, is an indicator of their relationship. But I must add that when I looked in the archives in Pisa at the lab notebooks the group kept during that project, I found it difficult to disentangle from those laconic entries who did what, and when. Fermi's innovations, in that more innocent age, did not go so far as to foresee the kind of self-protective bookkeeping forced upon labs in our time by the expectation that (as in the case of a researcher in the laboratory of David Baltimore) even the Secret Service might be called in by eager congressional investigators who doubt the trustworthiness of reported scientific results.

As we know, the invention of publicly assigning and distributing responsibility in works that have multiple authorship has been successful beyond all reasonable expectations. Records are being broken year by year as larger and larger groups launch themselves against harder and harder problems. A paper in the *Physical Review D* of 1 June 1992 lists some 365 authors from 33 institutions on 3 continents.[18] Such indicators signal a kind of phase change in the social matrix of science, to which we shall have to turn in closing. Before that, however, we must at least briefly note another crucial milestone on the road to greater reliability of findings: the adoption of statistical methods for the analysis of data.

The Statistical Treatment of Data

An example from the first decade of this century will underline how recent, from a historical perspective, are the details of data processing. These issues, such as the use of statistics and even of significant figures, are now so common that many assume they have been with us since the beginning of time. For this example I return to Robert A. Millikan, a scientist whose use of data in his unpublished lab notes I have had occasion to write about;[19] I found it a revealing case study of the way a creative researcher often exercises judgments during the nascent, private phase of experimental work that may look impermissibly arbitrary when examined with the benefit of hindsight. This time I want to look briefly at Millikan's publication of February 1910 in the prestigious *Philosophical Magazine*, his earliest "big" paper on the road to what two years later turned into his triumphant "oil drop" method for measuring the charge on the electron, e.[20]

In 1910, at age 42, Millikan was still essentially unknown, and he was near despair about his chances to break through to scientific prominence. He had no way of knowing that the February 1910 paper, on a "new modification" of a well-established method of using the motion of water droplets in electric fields to find e, was to point him soon in the right direction. But the paper allows us to glimpse how, at that point in early twentieth-century history, one could still treat one's data.

In the section entitled "Results," Millikan frankly begins by confessing to having eliminated all observations on seven of the water drops—these, for various reasons, he decided had to be "discarded." A typical comment of his, on three of the drops, was: "Although all of these observations gave values of e within 2 percent of the final mean, the uncertainties of the observations were such that I would have discarded them had they not agreed with the results of the other observations, and consequently I felt obliged to discard them as it was."[21] Today one would not treat data thus, and one would surely not speak about such a curious procedure so openly.

Millikan then presented his results in tables, in each of which he had gathered the data from one of six series of experiments, and gave also the raw calculations they yielded. Almost every detail there is astounding from our present point of view. Each observation carries Millikan's opinion of its likely reliability: "The observations marked with a triple star are those marked 'best' in my notebook....The double-starred observations were marked in my notebook 'very good.' Those marked with single stars were marked 'good' and the others 'fair.'"[22] Correspondingly, he assigns to each of the six series a weight to be used in averaging all results to obtain a final value; no details are given, but by inspection one sees that in Millikan's mind he correlates nine or ten stars in a series with a weight of 7, seven stars with a weight of 6, five stars with a weight of 4, three with 3, and zero with a weight of 1.

Within each of the six series in the tables, there are more surprises. There is little attempt to keep the significant figures straight during calculations; no reason is given why the readings are clumped the way they are, except that each clump corresponds to a set of drops that are guessed to carry the same number of electron charges. To find a value for e within a clump of observations, Millikan uses the average of all the individual data for the voltages needed to balance the various drops and of the times of fall when the electric field is off—instead of calculating the charges drop by drop and then applying statistical data reduction for a final result. Throughout, an air of utter self-confidence pervades the paper. But one must add one more fact: Despite all these "peculiarities," the final result obtained with the new method for the charge of the electron ($e = 4.65 \times$

10^{-10} e.s.u.) was excellent for its day, and could not be improved for many years. Oersted's phrase, "the anticipatory consonance with nature," comes again to mind.

Statistics had entered theory in the eighteenth century and experimental science by the mid-nineteenth century. But it took much longer to get general agreement on the proper use of statistical analysis—e.g., to deal with such bothersome problems as outriders in lab data, and when or how to "reject" observations; for in the real world of the laboratory, unlike its idealization by many nonscientists, one must be prepared to find even that "at some level, things will happen that we cannot understand, and for which we cannot make corrections, and these 'things' will cause data to appear where statistically no data should exist....The moral is, be aware and do not trust statistics in the tails of the distributions."[23] Only in the mid-twentieth century did good practice on those points become general at the level of the ordinary lab worker. When excellent books such as E. Bright Wilson's *Introduction to Scientific Research* (1952)[24] became available, every student could read what "reasonable procedures" for data acquisition and reduction might be, including under extreme or difficult conditions.

"Big Science"

But of course the landscape has changed immensely in the decades since then. Our journey brings us now to the latest and most difficult stage, that of the gigantic thought- and work-collectives in science. The story is told well in Peter Galison's book, *How Experiments End.*[25] Chapter 4 is devoted to the discovery in the 1970s of so-called neutral-current events, such as the observation of neutrino-electron scattering. The detection of these events led to the confirmation of the theory of the unification of the electromagnetic and weak nuclear forces by Glashow, Weinberg, and Salam—a modern triumph in the *Grundkraft* program of Kant and Oersted.

The theoretical questions and fierce debates concerning the possibility of such a unification took an entirely new turn when, at the big accelerator in CERN, a photograph of a single electron event was found by a group from Aachen in January 1973. The detecting instrument they used had been given the name Gargamelle, after the mother of Gargantua. It was a monstrous bubble-chamber device, holding 12 cubic meters of liquid propane—a major engineering project of its own, and a symbol of the interpenetration of engineering and science in our century. Thousands upon thousands of pictures are exposed in the course of an experiment and then are given to the team trained to scan the negatives. In this run, one exposure was recognized as being strange and new by one of the scanners. She passed it into the hands of a research student, who identified the traces as those of electrons. The following day the student carried it up along the complex hierarchy of a big group to the next rung on the ladder, to the deputy group leader. Thinking it of considerable interest, he in turn brought it to the institute director, who later called it a picture-book example of what they had been expecting for months: a candidate for neutrino-electron scattering.

The crucial point now was to assess the background, i.e., the probability of a masking event. The director took the picture to England, to a fellow expert. And so on, and so on. The circle of belief expanded constantly. But still, this was only one picture, a single event, and not even an event one could immediately reproduce. What a huge difference from the days of DuFay and his followers, all the way to Fermi's group! Yet, as Galison writes: "Experienced bubble chamber experimentalists found the Aachen electron particularly compelling. Their specialty was famous for several critical discoveries grounded on a few well-defined instances. The omega-minus [particle] had been accepted on the basis of one picture, as had the cascade zero. Emulsion and cloud chamber groups had also compiled arguments based on such 'golden events,' including the first strange particles, and a host of kaon decays."[26]

Still, too much was at stake to publish immediately. One might believe the evidence, but one could not yet believe *in* it, trust it to be reproducible in principle. One now had to calculate the probabilities of all kinds of other reactions that could masquerade as the one assumed to be happening. Background events had to be ruled out; if one of those was what the exposure was about, it wouldn't test the Glashow-Weinberg-Salam theory at all.

There ensued an agonizing set of discussions within the large collective, as drafts of publications were being debated. We see here the next step in the evolution we are tracing: the internalization within a large group of fellow workers of the array of old but public methods of arriving at trustworthy results—debates among fellow specialists, preparation of publications and refereeing, and all the other safeguards of standards of demonstration. It is as if a new, large organism were doing science within its own boundaries. Finally, Galison concludes, "the members of the collaboration persuaded themselves that they were looking at a real effect. So it was that no single argument drove the experiment to completion....It was a community that ultimately assembled the full argument."[27]

The brief publication of the final result in *Physics Letters* announcing the find had boiled down years of work into a few sentences and neat graphs—a far cry from the extensive documentation of Kepler and Boyle.[28] A heterogeneous group of subcollectives from Belgium, Britain, France, Germany, Italy, Switzerland, and the United States had come to the decision to stake their reputations on a new kind of physics. And it is worth noting that the new finding had been distributed to the network of fellow specialists long before the mailing of the 19 July 1973 issue of the journal, through prepublication "preprints." Over the past few decades preprints have become in some branches of science by far the preferred method of "publishing" and of keeping up with the "literature"; the Physics Department library at Harvard alone receives over 4,000 preprints per year from more than 500 universities and other institutions worldwide. These in turn are now made quickly available through an electronically accessible database specially designed for them. Here, too, we catch a glimpse of the immense rate of change in the way science is being done. The next frontier is already in view: A debate is in progress among physicists at large colliders whether "outsiders" should have access to as yet unpublished data, even before the preprint stage.[29]

But to return to the CERN result. The story did not quite end there. By November 1973, an American group at Fermilab had gathered enough evidence that all this work at CERN was a mistake—that no neutral current existed. They were almost ready to announce their finding. But then they found traces of the effect that "would not go away," and in April 1974 they too published their evidence in favor of the neutral current.[30]

One lesson of this story is that just as one can trace the influence of Bach and Beethoven in some of Schoenberg's music of the early twentieth century, so can one find some habits, methods, and standards of the old days in the mega-teams of modern science. But the new conglomeration is also a transformed entity, and it deals with a transformed science. In many branches, the more difficult it is to find credible access to the phenomenon, the more dependent the scientist is on apparatus built by others, on data gathered by still others, and on calculations carried through by yet others. When one is immersed in a large group, some members of which might change from month to month, it is not easy to know whom to believe, when to believe it, and how well to believe it. And the ground is shifting once again before our eyes. More than 900 physicists had signed on to a single experiment on the planned superconducting collider in Texas before its cancellation, and the number of collaborators on the Human Genome Project is greater still.[31]

All these changes raise a vexing question. Where lies the source of trust—for a participant in the research as well as for the reader of the publication—when most individuals within a team of widely diversified competences cannot vouch for every aspect of the published results, when perhaps not even one person in the whole group can be expected to be fully conversant with every element? A bitter rule of thumb has evolved: whereas credit for success tends to be parcelled out unequally (those close to the field usually feel they "know" who in a long list of authors had the best ideas or most novel techniques), an embarrassing error or misconduct by even a single participant is likely to bring discredit to all the members of the team. Thus there is now a big incentive to invent ways to distribute partial responsibility early and proportional justice, when necessary, later. For example, some research papers identify the contributions of individual contributors to a complex group effort. But because of the differences in location, or the wide spectrum of subspecialties involved, or the styles of leadership, etc., there is such a variety in the way teams are run that no single set of rules is likely to emerge. We are watching an experiment in self-governance, analogous perhaps in some ways to the transition that took place when the Pilgrims arrived in the New World and set about the task of forming a new society.

Doing One's Damnedest, Revisited

Still, some fundamental things apply as time goes by. It was my luck to have known, and to have learned my trade under, one of the last of the physicist-philosophers, P. W. Bridgman—a Nobel Prize winner (1946) for establishing the field of experimental high-pressure physics almost singlehandedly, and also the father of the movement in philosophy of science known as operationalism. Operationalism centers on the position that the meaning of a concept is in its measurement or other test. In the 1930s Bridgman made a famous, useful, and very operational statement, usually remembered as:

> The scientific method is doing your damnedest, no holds barred.[32]

For him personally, doing one's damnedest meant total absorption, from 8 a.m., when he arrived at the lab by bicycle, rain or shine. It meant incredible productivity: Bridgman published some 230 substantial scientific papers that came to seven volumes of his collected scientific works,[33] as well as several books on the philosophy of science. It meant unceasing dedication to the process of getting things clear in his own mind. His style was completely "hands on," every datum taken by him or by his one, dedicated long-term assistant. Most apparatus was built by his own hands.

Only two or three joint papers (one with a young chemist named J. B. Conant). Very few thesis students—only those whom he could not persuade otherwise. Only a few hundred dollars a year needed for materials. No overhead. No paperwork. Paradise!

New Demands on Scientists

Bridgman's statement is still a great definition of how to get at trustworthy scientific results. But in the intervening decades the milieu has so evolved that we need to supplement that advice, to bring it into our time. At the very least we must add a few words:

> While doing your damnedest, watch how your presuppositions are holding up; make sure you understand the results of other people along the chain on whose work you are relying; and keep in view that more and more of the findings of science, resonating through society, have additional results far from those sought initially.

Bridgman himself discovered that last point in his own way. He once thought that science is essentially "value free"; but with the rise of Nazism in the 1930s, he saw that German scientists were no longer able to act as free intellectuals, and were in many cases co-opted as servants of a Fascist state. Therefore he published a "Manifesto" in 1939, to the effect that he was closing his lab to visiting scientists from totalitarian countries.[34] The act served as an early, widely noted reminder for scientists in the U.S. that their work has ethical dimensions even beyond what Jacob Bronowski identified as the principle that binds society together, in science as well as outside, the "principle of truthfulness."

Most scientists today face not only new conditions, such as their complex relationships with sophisticated instruments and their collaborators. They also face a different kind of fact, one that increasingly asserts itself: By obtaining the necessary support from government or industry, they have new obligations, not least the responsibility that their results be sharable far beyond their own labs. For reasons having little to do with the relatively rare case of scientific error, misjudgment, or misbehavior, new pressures for accountability are jostling them from all directions. All parties involved are still fumbling around a bit—the NIH, the Congress, the universities, the scientific societies. They are struggling to invent new methods for preventing and defending against charges of misconduct, real or imagined—just as, at an earlier stage of evolution, they had to learn about factuality and significant figures.

This new task, too, is one which history has thrown their way. Scientists cannot let others create the new realities for them. The next phase in the continuing evolution and transformation of the methods of science will depend on the actions of today's researchers. At stake are the problem choices for tomorrow's scientists, their rights to respect and support, the attraction of their fields for future scientists, and even the ancient hope that scientific thought itself is an exemplar for attaining trustworthy conclusions.

References

1. See, e.g., Owen Gingerich, "'Crisis' versus Aesthetic in the Copernican Revolution," in Arthur Beer and K. Aa Strand, eds., *Copernicus. Yesterday and Today* (Oxford: Pergamon Press, 1975), pp. 85–93.
2. N. Copernicus, *De revolutionibus orbium coelestium* (Nuremberg: 1543), f. iii (v).
3. See Carl Seelig, *Albert Einstein* (Zurich: Europa Verlag, 1954), p. 195.
4. H. A. Lorentz, *The Einstein Theory of Relativity* (New York: Brentano's, 1920), pp. 23–24.
5. H. C. Oersted, "On the Spirit and Study of Universal Natural Philosophy" in *The Soul of Nature*, trans. Leonora and Joanna B. Horner (London, 1852; reprint London, 1966), p. 450.
6. Galileo Galilei, *Dialogues Concerning Two New Sciences*, trans. Henry Crew and Alfonso de Salvio (New York: Macmillan, 1914), p. 179.
7. Alexandre Koyré, *Galileo Studies*, trans. John Mepham (Atlantic Highlands, NJ: Humanities Press, 1978).
8. Galileo, p. 170.
9. Galileo, p. 153.
10. See Steven Shapin and Simon Schaffer, *Leviathan and the Air Pump: Hobbes, Boyle and the Experimental Life* (Princeton: Princeton University Press, 1989).
11. Hans Christian Oersted, "Experiments on the Effect of a Current of Electricity on the Magnetic Needle," *Annals of Philosophy* 16:273–76 (London: 1820). This is the English translation of Oersted's pamphlet, *Experimenta circa effectum conflictus electrici in acum magneticam*.
12. Oersted, pp. 273–4.
13. For some of these estimates, see G. L'E. Turner, "The Microscope as a Technical Frontier in Science," in *Historical Aspects of Microscopy*, ed. S. Bradbury and G. L'E. Turner (Cambridge: W. Heffer & Sons, 1967), pp. 175–97; on the astronomical angular measures, see H. T. Pledge, *Science Since 1500* (New York: Harper, 1959), p. 291.
14. See James Bryant Conant, ed., *Harvard Case Histories in Experimental Science. Case 1. Robert Boyle's Experiments in Pneumatics* (Cambridge: Harvard University Press, 1967), p. 62.
15. Dong-Won Kim, *The Emergence of the Cavendish School: An Early History of the Cavendish Laboratory, 1871–1900*, Ph.D. dissertation, Harvard University, 1991.
16. Sir James Chadwick, "Some Personal Notes on the Search for the Neutron," delivered at the 10th International Congress of History of Science at Cornell University, Ithaca, New York, in 1962; reprinted in *The Project Physics Course Reader* (New York and Toronto: Holt, Reinhart & Winston, 1971), Unit 6, p. 28.
17. E. Fermi, E. Amaldi, B. Pontecorvo, F. Rasetti and E. Segrè, *Ricerca Scientifica* 5 (2):282–283 (1934); translated and reprinted with the title "Influence of Hydrogenous Substances on the Radioactivity Produced with Neutrons—I," in Enrico Fermi, *Collected Papers*, Vol. 1

(Chicago: University of Chicago Press, 1962), pp. 761–2.

18. F. Abe, et al., "Limit on the top-quark mass from proton-antiproton collisions at $\sqrt{s} = 1.8$ TeV," *Physical Review D* 45, 3921 (1992).

19. See Gerald Holton, "Subelectrons, presuppositions, and the Millikan-Ehrenhaft dispute," in *The Scientific Imagination* (Cambridge: Cambridge University Press, 1978), pp. 25–83.

20. Robert A. Millikan, "A New Modification of the Cloud Method of Determining the Elementary Electric Charge and the Most Probable Value of that Charge," *Philosophical Magazine* 19:209-28 (1910).

21. Millikan, p. 220.

22. Millikan, pp. 221–3.

23. Philip R. Bevington and D. Keith Robinson, *Data Reduction and Error Analysis for the Physical Sciences*, 2nd ed. (New York: McGraw-Hill, 1992).

24. E. Bright Wilson, Jr., *An Introduction to Scientific Research* (New York: McGraw-Hill, 1952). Reissued in 1990 by Dover Publications Inc., New York.
See Theodore M. Porter, *The Rise of Statistical Thinking 1820–1900* (Princeton, NJ: Princeton University Press, 1986) for a guide to the development of the concern with statistics in science.

25. Peter Galison, *How Experiments End* (Chicago: University of Chicago Press, 1987).

26. Ibid., p. 183.

27. Ibid., pp. 193–4.

28. F. J. Hasert et al., "Observation of Neutrino-like Interactions without Muon or Electron in the Gargamalle Neutrino Experiment," *Physics Letters B* 46:138–40 (1973)

29. See, for instance, the letter from Oscar Hernandez in *Science* 258:13 (2 October 1992). On federal courts ordering the "sharing" of confidential data, see *Science* 261:284-285 (16 July 1993).

30. After accepting the existence of neutral currents, one of the chief investigators wrote: "Three pieces of evidence now in hand point to the distinct possibility that a [muon]less signal of order 10% is showing up in the data. At present I don't see how to make these effects go away." Galison, p. 235.

31. See Faye Flam, "Big Physics Provokes a Backlash," *Science* 257:1468 (11 September 1992).

32. P. W. Bridgman, *Reflections of a Physicist* (New York: Philosophical Library, 1955), p. 535. The passage from this source reads: "The scientific method, as far as it is a method, is nothing more than doing one's damnedest with one's mind, no holds barred."

33. P. W. Bridgman, *Collected Experimental Papers*, eds. Harvey Brooks, Francis Birch, Gerald Holton, and William Paul (Cambridge: Harvard University Press, 1964).

34. P. W. Bridgman, "'Manifesto' by a Scientist," *Science* 89: 179 (1939). For more discussion about Bridgman's style of work, see the chapter on Bridgman in this collection.

Imagination in Science

The arts and the sciences are usually thought to belong to two different worlds, but in some respects they are close cousins. For while the aims, tools, and products differ, the ingenuity and passion behind the two endeavors are similar. There is also a long history of mutual stimulation between art and science, ever since Pythagoras, who held that both music and the phenomena of nature are governed by relations among whole numbers. And as I shall try to show, historians of art have given us key insights into problems in the history of science.

But if we wish to watch the imagination of scientists at work, we shall have to catch them unawares. For quite good reasons, modern scientists try to keep their personal struggles out of their published research results and out of their textbooks. On that point, Louis Pasteur's advice to his students and colleagues still applies: "Make your results look inevitable." It is therefore chiefly through private records and laboratory notebooks that historians of science can find out what the scientists themselves do not care to reveal. For while logic, experimental skill, and mathematics are constant guides, they are by no means adequate to the task of scientific investigation—otherwise a computer could do original research unaided. When you listen at the keyhole of the laboratory door, you find that the scientist uses a variety of other tools as well. I shall share with you examples of three such tools in the doing of science, three closely related companions that are rarely acknowledged. I shall speak about the *visual imagination, the metaphoric imagination, and the thematic imagination.* My examples will be drawn mostly from physics, but one could harvest similar instances from other branches.

The Visual Imagination

One must start with the visual imagination, if only because early Western science made its debut through the eye—through watching the puzzling motions of the planets, the wanderers among the fixed stars. So it is not surprising that entities that could be imagined but were hidden from direct visibility were often regarded with great suspicion. For example, the elusive ether seemed a necessary base for understanding the propagation of light, which consists of transverse electromagnetic waves; but in order to replicate the supposed motions within such an ether one had to think up ever more fantastic mechanical models. Finally Heinrich Hertz cried a halt to it, saying that the mathematical equations describing light are all that we can imagine when we think of the motion of light waves.

Similarly, the ancient notion of the atom as a tiny, uncuttable, discontinuous entity became more and more inadequate as new electrical, chemical, and other properties of matter had to be explained. At the beginning of this century, scientists such as Ernst Mach threw up their hands at the very idea of the atom; he asked everyone sarcastically: "Have you ever *seen* one?"

In fact, it would not have been impossible to have some sort of physics and chemistry without postulating the existence of atoms (as the great chemist Wilhem Ostwald did in the first edition of his text, *Allgemeine Chemie*); but it would have been a much more complex, less beautiful science. Luckily, the eye came to the rescue. In 1912, the physicist C.T.R. Wilson showed these photographs (Figure 1) at a scientific meeting, and that settled the matter for most people. He had directed a beam of alpha particles from a radioactive source into a cloud chamber, a small glass box containing moist air at a low temperature. Along the path of each of the alpha particles, which are themselves of course invisible, there settles out a streak of fog, a thin cloud. That gives away the path of the alpha, somewhat as vapor trails in the sky are indirect evidence of the flight of an aircraft.

That was spectacular enough; but the real excitement was in the discontinuities of some tracks, their sudden changes of direction (e.g., the one shown in the lower left corner). There the alpha particle had hit something and scattered off in a new direction. In one case, the obstacle with which it had collided—namely a nucleus in one of the gas molecules—had been given enough of a push to leave its own tiny bump of a vapor trail. These pictures are simple, silent, and still; there is no evidence of motion. By itself, each is just a parsimonious hieroglyph. But to the properly prepared mind connected to an alert eye, the photographs presented an overwhelming drama—the first, irrefutable evidence of the existence

Figure 1.

of atomic discreteness far below the level of direct perception. Scattering of particle beams became *the* way to "see" subatomic events.

The next generation of tools to glimpse such events was the bubble chamber. There the tracks are traced out in a liquid and become visible as rows of tiny bubbles. Figure 2 is a famous example. The photograph looks again more or less like graffiti, but one learns to neglect in this case the unimportant scratches and curlicues and to focus only on five lines. They

Figure 2.

reveal that a life-cycle story had taken place on this tiny stage, as drawn in Figure 3: a pion—an elementary particle whose track is marked on the drawing that interprets the raw observations in Figure 2—came into the view field from the bottom edge. It encountered an unsuspecting proton in the chamber and interacted with it to form two so-called strange particles (labeled K^0 and Λ^0—called "strange" because they survived unexpectedly long for such created particles, namely of the order of one ten-billionth of a second). But these, being neutral, left no traces to look at, until they too decayed. The result of each "strange" particle's decay was one negative and one positive particle, producing in our view field, as it were, a third generation, each again having its own characteristic lifetime.

You notice that in such a technical description the physicist is using the rhetoric of a familiar, primordial type of drama or folk tale, acted out in

Figure 3.

space and time; a story of birth, adventure, and death. This anticipates a point to be made later: the power of many useful scientific concepts rests at least in part on the fact that they are anthropomorphic projections from the world of human affairs, and thus rest on metaphors.

Next let us look at another bubble-chamber photograph, one first noticed in 1973. By that time the bubble chamber had grown into a monster containing twelve cubic meters of liquid propane and was nicknamed Gargamelle, after the mother of Gangantua. Among the thousands upon thousands of pictures taken at CERN, where invisible, accelerator-created neutrino particles were beamed into the bubble chamber, one of the scanners noticed the frame shown in Figure 4, a photograph unlike any other. On analysis, what she had found turned out to be what is referred to as a "golden event"—a glimpse at a rare but very revealing interaction.

The naive eye must be told to neglect almost everything in that photograph and to focus on the limp curlicue on the left; it is a typical signature of an electron. The interpretation of this event helped finally to confirm the theory of the unification of the electromagnetic and the weak force, called the electroweak force. For that achievement, Sheldon Glashow and Steven Weinberg in the U.S. and Abdus Salam at Trieste shared a Nobel Prize. I shall come back to this picture shortly to indicate how the scientific imagination dealt with it.

Before we are ready for that exercise, we must go back to the beginning of modern science in the seventeenth century, to understand better the immense power of the iconic imagination—the ability to form suc-

Figure 4.

cessful mental pictures out of elusive optical images and so to convert vague perceptions into solid knowledge. My example of this process of conversion from optical to mental imagery concerns Galileo Galilei, in a case investigated by the art historian Samuel Edgerton, whose extensive analysis I shall outline.

This is the story. In 1609, two men looked at our moon through a new invention, the telescope. The first one to do so was the mathematician, cartographer, and astronomer Thomas Hariot in London, who used a 6-powered telescope, starting in late July 1609. The other was Galileo, then a professor of mathematics at the University of Padua; he had taught himself how to grind lenses and made a 20-powered telescope, with which he observed the moon in late autumn of the same year. Luckily we have a record of what each of these two men thought he saw. It is instructive to compare their private notes, and to understand the reasons for the great differences between them.

Figure 5.

Both men knew that at least since Aristotle the moon was widely thought to be a perfectly smooth, unchanging sphere, the symbol for the incorruptible universe beyond earth. Also, in paintings since the Middle Ages the moon had been a sign of the Immaculate Conception of the Virgin Mary; Figure 5 (from the painting by Bartolomé Murillo) is an example. But there were two problems with the "perfect" moon. One was that some areas on the real moon are obviously darker than others, so it couldn't be uniform throughout. Thomas Hariot called it "that strange spottedness." The second problem was that if the moon were truly a mirror in the shape of a perfect sphere, it would reflect to us at any time the image of the sun on only one small region of it, like a brilliant spot on a ball bearing, leaving the rest dark. But as always, suitable *ad hoc* theories sprung up to deal with these problems. For example, some said that the moon's surface was etherial, or like alabaster it re-emitted the light in a diffused way, allowing different internal materials to show through.

Hariot's first observation has been preserved among his papers (Figure

Figure 6.

6). It is a rough sketch showing the "terminator," the divider between the dark, night areas and the brightly illuminated day portion of the moon. But the important thing to note is that evidently Hariot doesn't know, and he makes no comment on, why it is a *jagged* line instead of the smoothly curved one which one would expect if the moon were indeed a perfect sphere. He *sees*, but the current theories of the moon's perfection make it difficult for him to understand what he sees.

Enter Galileo. Starting in late November 1609, he carefully scans the ghostly moon through his telescope and makes several skilled sepia drawings of his observations (Figure 7). Galileo evidently also sees the jagged lines along the terminator. But he interprets them at once as irregularities of the surface, as mountains and craters, and he uses the chiaroscuro technique of drawing and painting to manipulate dark and light, so as to emphasize the protuberances and depressions.

What Galileo had seen was soon beautifully described in his book of 1610, the *Siderius Nuncius*. Figure 8 shows one of the illustrations in the book; it exaggerates the moonscape for further effect. Galileo writes that the surface of the moon, contrary to current philosophy, "is not smooth, uniform, and precisely spherical...but uneven, rough, and full of cavities and prominences, not unlike the face of the earth, relieved by chains of mountains and deep valleys." Galileo *sees* that there is no qualitative difference between earth and moon. He even calculates from the shadows

Figure 7.

cast by the peaks that the mountains must be four miles high from the baseline—higher than the Alps on earth! His voice is calm; but he knows that the ancient Aristotelian world view is crumbling under his blows.

The news of Galileo's sensational findings spread throughout Europe and transformed what people saw—an example of how the meaning conveyed by objective data depends on the presuppositions held by the viewer. Thomas Hariot himself, having read Galileo's book, raised his telescope again in July 1610, just a year after his first try, and made a sketch of his new observation (Figure 9). Now he too sees mountains and shaded craters—even more of them than were in Galileo's published sketch. Having been converted to a new way of looking, having abandoned his old presuppositions, he now sees something quite different in the same old moon. I am reminded here of that wonderful passage in Tolstoy's *Anna Karenina*, where Anna, hopelessly in love with Count Vronsky, explains to a friend that she cannot love a man like her husband Karenin, because Karenin has such big ears. To this the friend wisely replies that what has changed is not the ears of Anna's husband but the heart of Anna.

Now we must ask what it was, before Hariot's change of heart, that gave Galileo and Hariot such different eyes initially, when they looked at the same object. Part of the answer lies of course in Galileo's greater

Figure 8.

readiness to consider a Copernican universe, in which planets and satel-
lites can all be similar. But much of the answer lies also in their training in
visualization, in how they had learned to use their eyes as a tool of imagination. In Hariot's England of 1609, the peak of artistic achievement was
the *word*, for example that of Shakespeare, rather than anything in the visual arts. Indeed, visually England was far behind—one might almost say
in the Middle Ages—with respect to understanding perspective renditions.
In Galileo's Italy, however, Renaissance painting had captured the alert
intellectuals. In 1562, under Cosimo I of Florence, Vasari had founded the
great Academy of Design, a center for visual arts and architecture for the
benefit of all, not only the practitioners. It is not an accident that Galileo's
first job application, made at age twenty-five, was for a position of professor
of mathematics at the Academy, to teach geometry and perspective. And in
1613 he was elected to membership of that distinguished Academy.

So it is very likely that Galileo, like all the students at that Academy,
had studied the problem of how bodies cast shadows on different surfaces. The typical, well-thumbed texts used in the Academy show how the
raised protuberances and depressions on reticulated spheres appear in
light and shade (Figure 10). The art of perspective rendering in chiaroscuro was a tool and an ability which Galileo had learned as a young man.
It came suddenly to good use when the old shadow-casting problems re-

Figure 9.

appeared before his eyes in 1609 in the entirely different context of the
telescopic view-field. One may say he was able to *envisage*, through this
still rather poor optical tube, how scientists everywhere would soon come
to see and understand the phenomena in the solar system.

This case was an exemplar of that powerful mix in doing science: hard
data plus solid skills of mathematics and praxis, plus theoretical preconceptions, all working together in the theater of the mind. And in this mix,
the skill to use the visual imagination has often been crucial. In a previously quoted letter to Jacques Hadamard, Einstein confessed as follows:
"The words or language, as they are written or spoken, do not seem to
play any role in my mechanism of thought. The psychical entities which
seem to serve as elements in thought are certain signs and more or less
clear images which can be voluntarily produced and combined." It was as
if in thought he played with the pieces of a jigsaw puzzle. And to Max
Wertheimer, Einstein reported, "I very rarely think in words at all...I have
it in a sort of survey, in a way visually."

That was of course why Einstein, as a young man in Bern, had been a
good Patent Office examiner. His job was to study the descriptions, and
particularly the drawings, submitted by the inventors and to reconstruct
those proposed machines in his mind to see whether they could really function. That ability came easily to him. And in his physics, too, he could visualize without effort processes which others found uncomfortably complex.

Let me give you a simple example of visualization. If you have studied

Figure 10.

physics as far as the introduction of special relativity theory, your text-book no doubt asked you to suppose that a train is going very fast past a station platform on a stormy day. You had to imagine that one observer is standing on the platform and the other is traveling in the middle of the train. Suddenly two lightning bolts come down from the sky, each hitting the moving train; one hits in front, and the other one in back. The important question was: How will these events look to each of the two observers, the one who is stationary on the platform and the other traveling at high speed?

You will remember what the answer was: if to the first observer the two lightning bolts appeared to crash down simultaneously, to the other one (who is moving toward one of the flashes and away from the other), they will appear to have been separated in time. This proves that the si-multaneity of two events is not absolute for everyone but depends on the state of motion of the observers; it is "relative."

A lot of physics follows directly from visualizing this whole scene in your thoughts, from correctly doing this "thought experiment." And that visually dramatic example really does come directly from Einstein's own writings. (A diagram in his 1917 book on relativity has his typically parsi-monious sketch of the situation.) All this was child's play for him, but not so easy for others. It took them time to learn how to see, to imagine it.

Nowadays it has become perhaps too easy. Einstein's imagery has found its way even onto the theater stage. In the five-hour opera by Robert Wilson and Philip Glass with the title "Einstein on the Beach," there is a prominent depiction of that train; but in the opera it is creeping across the stage very slowly for two long acts, and something that represents the lightning bolt is also moving very slowly from above. Einstein would have been amazed to see this, since in his example everything depended on the train moving very fast.

At any rate, Einstein's visual imagination served him superbly again and again. Some time ago I found in the Einstein Archives one of his manuscripts from about 1920, in which he tells how he came to invent the *general* theory of relativity. As noted earlier in this book, the key here was to realize that the effects of accelerated motion and of gravity can be considered equivalent. As Einstein described it in the manuscript: One day in 1907 "there came to me the happiest thought of my life," namely that "the gravitational field has only a relative existence. For if one considers an observer who is falling freely from the roof of his house, there exists for him during his fall no gravitational field." For example, any object he releases while falling will just stay near him. That was a breathtakingly simple, visualizable scientific "thought experiment," the basis of the equivalence principle of general relativity.

In the early part of this century, the iconic imagination continued to lead from one scientific triumph to another. For example, Niels Bohr's atomic model of 1913 adopted the imagery of the Copernican solar system. It was of course a great breakthrough. But by the mid-1920s it became clear how dangerous it was to think about atomic processes in terms originally invented for large-scale events such as the motion of planets. A new way was needed to imagine phenomena such as the "spin" of the electron, or light being considered both as a wave and a particle. The easily visualized, model-based intuitions, as opposed to conceptual abstraction, had become an obstacle. One does not need to know much about Heisenberg's uncertainty principle to realize that those precisely drawn orbits in Bohr's atomic models really cannot exist in nature.

This principle led Heisenberg from 1925 on to propose a necessary but drastic solution, one which to this day makes it so difficult for laymen to feel at home in modern physics. Heisenberg totally eliminated the use of picturable models of the atom. A typical Heisenberg dictum was: "The program of quantum mechanics has to free itself first of all from these intuitive pictures...The new theory ought, above all, to give up visualizability totally." Or as P.A.M. Dirac wrote in 1930: "The classical tradition has been to consider the world to be an association of observable objects...It has become increasingly evident in recent times, however, that nature works on a different plan. Her fundamental laws do not govern the world as it appears in our mental picture in any very direct way, but instead they control a substratum of which we cannot form a mental picture without introducing irrelevancies."

In most other sciences today, the old iconic imagination lives on, alive and well. But the quantum scientists of today have gained a new kind of visualizability, though largely through mathematical rather than physical constructs, through symmetries and abstract diagrams. In Figure 11 we have at least a hint how the new way of thinking differs from the older one. At the top is pictured the traditional visceral picture one employed to talk about what happens when two equally charged electrons are brought near each other. It is a kind of snapshot of a situation in space; the two electrons mutually exert forces of repulsion somehow across the gap between them. But it is now found much more meaningful to think of this phenomenon as caused by the mutual exchange between these two particles of a photon, an entity that mediates the interaction. The lower part of the figure represents this new way of "picturing," by means of a kind of

Figure 11.

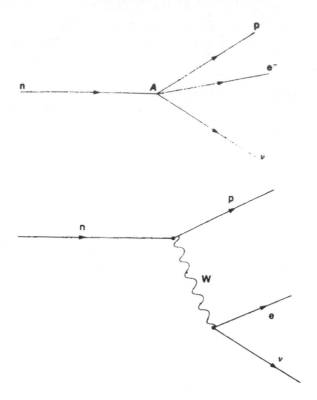

Figure 12.

diagram (named after its inventor, Richard Feynman) that provides a representation in space-time of the scattering of two electrons.

Similarly for the beta-decay of the neutron, which was first explained by Enrico Fermi. In the old way of picturing beta-decay (top of Figure 12), the interaction between the original neutron and the resulting proton, electron, and neutrino, everything takes place at a single space-time point A. By contrast, and as diagrammed in the lower half of Figure 12, the modern understanding of beta-decay imagines the interaction between the particles to be "spread out" in space-time and mediated by the exchange of a W-boson. The two presentations give the same predictions at low energies, but the results are quite different at high energies.

Thus as the simple mental models have withered away, new diagrammatic helpers for our thought processes have taken their place—diagrams in which each part stands for a mathematical expression needed to calculate forces of scattering probabilities. Figure 13 is another example. As my colleague Howard Georgi described it: "An important test of the modern theory of the electroweak force is the existence of neutral currents. This means that weak [i.e., rare, improbable] processes of the type shown

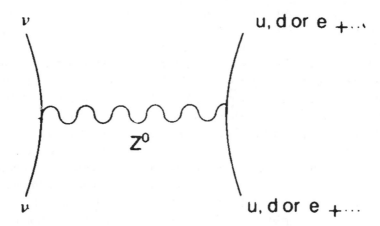

Figure 13.

can occur, in which an electrically neutral virtual quantum (Z^0) is exchanged between a neutrino [represented by the curved line on the left] and a quark [curved line on the right], leaving their identities [e.g., their charges] unchanged."[1]

And this brings us back, as I promised, to the "golden event" I spoke of earlier. For this quote is precisely the description of the underlying text of what happened in the photograph that was given in Figure 4. Our "naked eyes" would see only an unconvincing curlicue in the bubble-chamber picture. But the mind's eye sees, through the use of a Feynman-diagram version of the same phenomenon, that a neutrino scatters an electron without any change of charge; therefore a "neutral current" *exists*; therefore the electroweak theory is right; therefore if you happen to be Glashow, Weinberg, or Salam, who thought of this possibility first, you will have reason to be very satisfied with this experimental result.

The Metaphoric Imagination

Let us now watch another conceptual tool at work, one that some scientists use with great mastery in the genesis of their ideas. It is the *metaphor* and its close cousin, the *analogy*.

This might surprise you. After all, some philosophers have gone on record to say that the metaphoric imagination can have no good purpose in science. The *Dictionary of Modern Thought* still says of metaphor and analogy that they are "a form of reasoning that is particularly liable to yield false conclusions from true premises." Metaphor has been called the "essence of poetry;" it works through illusion. And surely, the business of scientists is precisely the opposite. Metaphor and analogy might therefore seem to be what scientists should most assiduously *avoid*.

Nevertheless, scientists do use analogies quietly all the time. The nineteenth-century physicist Thomas Young is a good example of how one can get chastised if one does it openly. His fame rests chiefly on arguing for the idea that light is fundamentally a wave phenomenon, contrary to the quasi-corpuscular theory that was widely preferred at the time. In one of his first published papers, Young writes: "Light is a propagation of an impulse communicated to [the] aether by luminous bodies." He reminds his reader that "It has already been conjectured by Euler that the colors of light consist of the different frequency of the vibrations of the luminous aether." But this has so far been only a speculation. Now, Young say, he has *confirmation*: The idea that light is a propagation of an impulse in the aether "is strongly confirmed..." (by what? how?) "by the *analogy* between the *colors* of a thin plate and the *sounds* of a series of organ pipes" (two entirely different phenomena).[2]

Even without stopping to study the details of this curious and, as it turned out, very fruitful analogy between light and sound—this surprising extension of the metaphor of wave motion from one field to another, seemingly quite unrelated one—we sense the remarkable daring of this transference of meaning. Indeed, making this connection, and so launching on the experimental proof of the wave nature of light, seemed very ill-advised even to George Peacock, a devoted friend of Young, and himself a scientist at Trinity College, Cambridge. When Peacock published a gathering of Young's articles in 1855—twenty-six years after Young had died, and long after the firm establishment of the wave theory—Peacock still felt he must save his reader from some dreadful mistake on this point; and so he added an asterisk after the crucial sentence just quoted from Young's paper and provided a stern footnote that is perhaps unique in the literature: "This analogy is fanciful and altogether unfounded. Note by the Editor."

Thomas Young's case is an exemplar of the creative but risky function of metaphor or analogy during the nascent phase of the scientific imagination. For Enrico Fermi it was part of his scientific credo to use and reuse the same idea in quite different settings. To him, any physical phenomenon could be understood in terms of an analogy with one of only about a dozen primitive, primary physical situations. For example, he effectively launched modern elementary particle physics with a paper on beta-decay in 1934, in which he said the puzzling emission from a nucleus of low-weight particles, such as electrons, should be understood by analogy with the well-established theory of the emission of light quanta (photons) from a decaying atom. In this way he avoided the trap of having to think of the electron as existing in the nucleus already before its emission; after all, no one had felt a need to think of the photon as already existing full-blown in the atom before it is radiated away.

And again, soon after writing a paper dealing with the effect that slow *electrons* have on colliding with an atom, Fermi was uniquely able to understand the effect of slow *neutrons* on the nucleus. This happened in October 1934, when he and his team, largely by accident, came upon the miraculously enhanced artificial radioactivity of silver, which turned out to have been caused by scattered, that is slowed-down, neutrons. The laboratory notebook pages that recorded this discovery are quite laconic, and the resulting paper was very short, less than two pages. But one could say that his use of analogy launched Fermi on what turned out to be the necessary first step toward the nuclear reactor, and hence to the so-called nuclear age.

The Thematic Imagination

With this example, I come to the third type of imaginative tools used by scientists during the nascent phase of research. I call it the thematic imagination. It is even more risky than those discussed so far: I am referring to the practice quietly to let one's presuppositions act for a time as a guide in one's own research when there is not yet good proof for them, and sometime even in the face of seemingly contrary evidence. This can amount to a willing suspension of disbelief, the very opposite of what one usually takes to be the skeptical scientific attitude.

Indeed, the phase "willing suspension of disbelief" comes from a discussion of poetry in Samuel Taylor Coleridge's *Biographia Literaria*. He saw his task, as he put it, to imbue his poetic writings with "a semblance of truth sufficient to procure for these shadows of the imagination that *willing suspension of disbelief for the moment* which constitutes poetic faith."

On the face of it, this surely can have nothing whatever to do with science. On the authority of the philosopher of science Karl Popper, we are told that the demarcation criterion of all truly scientific activities is the suspension of belief, not of disbelief. According to Popper, we must subject our rational constructs to a curative purging regime so as to look for a fatal flaw even in our own most treasured brain children. We must try hard to falsify, i.e., disprove them, and therefore to disown them.

And yet, when we stand there at the keyhole of the laboratory door, we observe many of our scientists paying no heed to this well-meant advice. Indeed, sometimes they let their best work grow and mature out of an unlikely idea that they prevent from being destroyed at the hand of iron rationality. Of course, eventually, after this private and nascent phase is over, the results obtained with maturer technique and the guide of maturer theory must stand up to experimental check. Nature cannot be fooled. The graveyard of science is crowded with the victims of some obstinate belief in an idea that proved unworthy. But we must face the strange fact that there *are* genial spirits who can take the risk, and persevere for long periods without the comfort of confirmatory support, and survive to collect their prizes. By studying their private notes we now know that Isaac Newton, John Dalton, and Gregor Mendel, among many others, refused to accept "data" that contradicted their presuppositions and were proven right in the end.

But the adoption of ardently held themata, and the suspension of disbelief in them, while necessary at some points and often successful, *can* ultimately mislead terribly. And to conclude with an example of a failure after having spoken of so many scientific successes, let me return to Galileo, and to a long-standing mystery about one of his few but grand errors.

As we all know, the climax of the scientific revolution for the physical sciences in the seventeenth century was Newton's *Principia*, which combined the imaginative breakthroughs of Galileo Galilei and Johannes Kepler. As Newton confessed, he saw further than others because he had raised himself onto the shoulders of giants such as these. Kepler, at the court of the mad and magnificent emperor Rudolf II in Prague, and Galileo, in Venice and Florence, were two very different personalities; but they had a great deal in common, above all their passionate devotion to the Copernican theory of the planetary system. Each braved the dangers which their heretical notions entailed, and Kepler, the younger by some eight years, an extravagant admirer of Galileo, tried in every way to get his attention and moral support.

Now it would have been very logical for Galileo to reciprocate, because Kepler's laws clearly showed the superiority of the Copernican way of imagining the system of the world. But contrary to every reasonable expectation, Galileo kept his distance from Kepler, always tried to brush him off, and never accepted Kepler's laws of planetary motion. And that, for a long time, was a maddening puzzle in the history of science. How could Galileo avoid using Kepler's supportive findings as a weapon when he was so beleaguered by his enemies? What caused this failure of imagination, one of the very few in Galileo's splendid *opera omnia*? He never tried to explain his strange refusal, and even this shows that it must have had a deep-seated cause. As the historian of science Giorgio de Santillana once said, the ideas of Kepler "must have set in motion a protective mechanism in [Galileo's] mind." What did he want to protect?

The explanation was finally found in a most unexpected way—again by an art historian, the magisterial Erwin Panofsky.[3] His brilliant analysis started from the fact to which I alluded earlier, that Galileo, like so many Italian intellectuals at the time, rightly regarded himself not only as a scientist but also as an admirer and critic of the arts. Moreover, to Galileo a chief criterion for sound scientific thinking was to use in his science only those thought elements that passed muster on aesthetic grounds. And it was on aesthetic grounds that Galileo found Kepler's ideas unacceptable and even repulsive.

Let me develop Panofsky's argument. Galileo, the son of a renowned musician and theorist of music, grew up in a humanistic rather than a scientific environment. It is well known, for example, that he devoted many months of patient labor to a comparison between the poets Oriosto and Tasso, extolling the former and tearing the latter to pieces. Galileo also threw himself gladly into controversies in the visual arts. For example, he was very close to Lodovico Cardi, called Cigoli, considered perhaps the most important Florentine painter of his time. In fact, Cigoli even collaborated with Galileo in astronomical observations; he also called Galileo his "master" in the art of perspective drawing. Cigoli indicated his admiration for Galileo when, in his last work, the frescos in Santa Maria Maggiore in Rome, he showed the ascending Virgin Mary standing on a moon that matched exactly what Galileo had recorded in one of his illustrations in the *Siderius Nuncius* (Figures 14 and 15).

In June 1612, Cigoli asked Galileo for help against those who were arguing that sculpture was superior to painting. Strangely enough, a key to Galileo's rejection of Keplerian astronomy can be found in Galileo's resulting letter on the superiority of painting. The trouble with sculpture, Galileo says, is that it is too closely akin to the "natural things," the objects with which it shares "the quality of three-dimensionality." The painter deserves greater credit for his work precisely because he has available only two dimensions to create the appearance of three-dimensionality. For, as Galileo put it, "The farther removed the *means* of imitation are from the *thing to be imitated*, the more admirable the imitation will be." And he adds for emphasis that we admire a musician if he "moves us to

Figure 14.

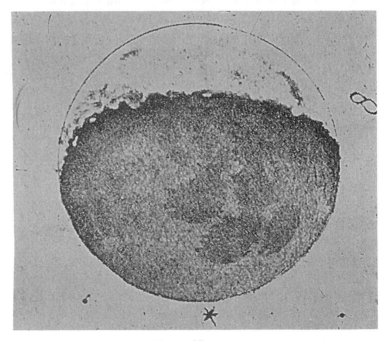

Figure 15.

sympathy with a lover by representing his sorrows and passions in song," but not if the musician simply just cries and sobs; and we should admire this musician even more if he didn't sing at all but used only musical instruments to act on our emotions.[4]

Galileo's point is that we must adhere to "critical purism;" we must keep the *representation* distinct from the subject matter. It is the same sharp knife Galileo used to separate quantity from quality, and science from religion. As Panofsky put it, Galileo objected to any blurring of borderlines. That is precisely why Galileo found Tasso's fanciful allegories so distasteful (for example in the poem "Gerusalemme Liberata"); and above all why Galileo, like Cigoli, also was opposed to artistic distortions that demeaned the medium of painting, as in "trick pictures." Galileo was particularly scathing of the then widely admired Giuseppe Arcimboldo, court painter of Rudolf II (to make things worse), a painter who had specialized in personifying concepts or seasons by arrangements of implements or of fruits and flowers (see Figure 16, representing Summer). This style, now referred to as Mannerism, arose as an "anti-classic" tendency, which, Panofsky pointed out, stood opposed "to the ideals of rationality..., simplicity, and balance" and instead favored "a taste for the irrational, the fanciful, the complex and the dissonant".[5]

Now there is one element in particular that was as emphatically rejected by high Renaissance art, which Galileo adored, as it was cherished in Mannerism, which Galileo abominated. And that is the ellipse. In painting and sculpture it was introduced as a significant element by Correggio and Gian Maria Falconetto, respectively; in architecture, Michelangelo had briefly toyed with the idea, in a design for the tomb of Pope Julius II, but it was to be only an interior feature, safely invisible from outside. For Galileo, the fight against Mannerism, against unnecessary complexity, distortion, imbalance, was a serious duty whether he dealt with music or poetry or painting.

And now we are ready to ask, with Panofsky, "If Galileo's scientific attitude is held to have influenced his aesthetic judgment, might not his aesthetic attitude have influenced his scientific theories?" More specifically, could it be that "both as a scientist and as a critic of the arts [he obeyed] the same controlling tendencies?"[6] We begin to see why Galileo thought Kepler was entirely on the wrong track. On the most obvious level, Kepler's writings, as in his *Mysterium Cosmographicum* and the *Harmonice Mundi*, are such uncontrolled outpourings of different ideas and subjects that it is difficult to see what is valuable under all this seeming fantasy. The three laws of planetary motion of Kepler, without which Newton could not have succeeded, are so deeply buried under mountains of debris that even Newton found it difficult to acknowledge his indebtedness explicitly.

Figure 16.

But quite apart from the indigestibility of Kepler's style of *writing*, Kepler's style of *thinking* seemed to Galileo to enthrone Mannerism in the solar system. To Galileo, as to Aristotle and also to Copernicus, all motion in the heavens had to proceed in terms of the superposition of circles, for example in a circular epicycle carried along a circular deferent. The circle, and uniform motion along the circle, were the very signatures of uniformity, perfection, eternity. Kepler had initially thought so, too, but then he had been driven by the data and against his better instincts to announce, as his first law, that the planets are in elliptical motions around the sun. Therefore they were not in what Galileo regarded as "natural" motion but were continually changing their speed as they moved.

So to Galileo, who was still completely under the spell of circularity, the ellipse was a distorted circle—a form unworthy of celestial bodies. To accept such an abomination was to give the victory to the Correggios and Arcimboldos of this world. That he would never do. Rather, the primacy of the circle was to Galileo what I have called one of those irresistible thematic presuppositions, without which his scientific imagination could not operate. And not only in the sky but on earth, too. As Galileo put it, "All human or animal movements are circular." Running, jumping, walking, and so on, are, as he says, only secondary movements depending on the primary one, which takes place at the joints; "it is from the bending of the leg at the knee, and of the thigh at the hip, which are circular movements, that the jump or the run results."[7]

The enchantment with the circle did not, in the end, undermine Galileo's cosmology in any serious way. But it did harm his physics, for it prevented him from coming to the realization that rectilinear rather than circular movement is the most natural one. Instead, Galileo held—as he put it in Book 1 of the *Dialogo*—that Nature allows straight-line motion only temporarily and only for the reestablishment of an order that has been disturbed. Once a piece of matter has reached its proper place, "it has to rest immovable or, if movable, to move only circularly." So Galileo missed the insight that is the very basis of modern mechanics and that we now refer to as Newton's first law, namely that in the absence of forces, bodies proceed with uniform speed in a straight line. It is truly ironic that the honor of discovering this principle of inertia had to go the Englishman, who by no stretch of the imagination could have considered himself an admirer or a critic of any of the arts.

We have now watched how some of the best scientists avail themselves of three of the chief tools of the scientific imagination. These examples show again what a caricature is the more common notion of scientific thinking as an almost irresistible, machine-like process of induction. Historians of science and other scholars the world over have been piecing together this more complex and chaotic, but more realistic and interesting understanding of how scientists have used their minds as they took on harder and harder questions during the past four centuries.

But I wish to finish on a cautionary note. By naming the varieties of imagination, we of course shall not have "explained" a Galileo or a Fermi, any more than a Mozart or a Verdi. We shall never fully solve the puzzle *how* some gifted scientists gain their foreknowledge of the coming state of science, how it is possible for our minds to discover the order of things. On this point, Albert Einstein again has to have the last word: "Here lies the sense of wonder, which increases ever more—precisely as the development of knowledge itself increases."

References

1. Howard M. Georgi, "Grand Unified Theories," in Paul Davies, ed., *The New Physics* (Cam-

bridge: Cambridge University Press, 1989), p. 435.
2. Thomas Young, "Outline of Experiments and Inquiries Respecting Sound and Light," in George Peacock, ed., *Miscellaneous Works of the Late Thomas Young* (London: John Murray, 1855), pp. 80–81.
3. Erwin Panofsky, "Galileo as a Critic of the Arts: Aesthetic Attitude and Scientific Thought," *Isis*, vol 47 (1956), pp. 3–15.
4. Quoted in Panofsky, p. 5.
5. Panofsky, p. 7.
6. Panofsky, p. 10.
7. Quoted in Panofsky, p. 13.

Further Reading

1. Samuel Y. Edgerton, Jr., "Galileo, Florentine 'Disengno,' and the 'Strange Spottednesse' of the Moon," *Art History* (Fall, 1984), pp. 225–232.
2. Peter Galison, *How Experiments End* (Chicago, IL: University of Chicago Press, 1987).
3. Jacques Hadamard, *The Psychology of Invention in the Mathematical Field* (New York, NY: Dover Publications, 1954).
4. Gerald Holton, "Metaphors in Science and Education," in *The Advancement of Science, and its Burdens* (Cambridge, UK: Cambridge University Press, 1986), pp. 229–252.
5. Gerald Holton, "The Thematic Imagination in Science," pp. 31–52, and "Thematic and Stylistic Interdependence," pp. 75–98, in *Thematic Origins of Scientific Thought: Kepler to Einstein* (Cambridge, MA: Harvard University Press, 1988 rev. ed.).
6. Arthur I. Miller, *Imagery in Scientific Thought: Creating 20th-Century Physics* (Boston, MA: Birkhäuser, 1984).

Understanding the
History of Science

W hat does it mean to "understand" the history of science—or at least to understand the history of *a* science? To prepare for an answer, let us think of what might be meant by saying that somebody understands a science, for example, physics, and see how that differs from understanding its history. On that question we have some guidance. Albert Einstein, for example, thought that the highest task of a scientist is to achieve an integrated picture of the physical world (a *Weltbild*). This can be understood in terms of an analogy: If one stands on a high mountain, one sees at a glance the whole varied landscape below, and especially evident are connections in the topography that would be hidden if one looked about from within one of the valleys. Similarly, from an integrated scientific view of nature one should be able to deduce in principle, and in this sense "understand," the detailed phenomena of nature. Classical mechanics is an example that fits this model, and the history of the development of science points toward the same goal generally.

We are of course a long way from attaining such a synoptic understanding. Still, a few scientists do have a remarkable understanding of this sort at least in their own branch of science; Enrico Fermi was an example. Even when a person of that caliber does not know fairly quickly how to solve a particular problem, he or she will make a good start on it—and is likely to go more or less in the same direction as another scientist of that level. This is probably because of four conditions: First, a problem in one of the exact sciences usually turns out to have one right answer. (That is of course very different from the situation in the history of science.) Next, most such problems are understood in about the same way everywhere; only rarely are there fundamentally different schools of thought in the exact sciences. (Again, this is quite different in the history of science.) Third, most experts in a given field share more or less the same scientific epistemology and ideology. And lastly, the raw material, the database for most problems in an exact science is or can be made relatively certain, because it can usually be reproduced or at least reexamined at will in a suitably equipped laboratory. (These last two conditions also do not apply to history of science. For example, a documentary base is much less easily sharable.)

So the first condition for understanding the history of science—and I shall include in it the history of mathematics, technology, and medicine—is that one cannot generally agree upon procedures or answers. Instead, there are markedly different, competing schools of thought. This has been true since the early period in the field. For example, George Sarton and his followers believed that the main task is the encyclopedic accumulation of precise knowledge of all innovators, throughout all history, and throughout all branches of science and technology. Joseph Needham and his school, on the other hand, approached the history of science through the development of science and technology of one great civilization (China), from its earliest beginnings. Herbert Butterfield felt that the important task is to understand the history of one great period—the rise of science in the seventeenth century. To others, history of science is chiefly a tool for supplying the "proof" of an ideological or epistemological position (e.g., to find demarcations between "progressive" and "degenerating" research programs). Still others use the history of science to illustrate their theories of the sociology of science (e.g., social construction), or to develop an evolutionary model of scientific progress, and yet another school of thought uses—often the same—cases to prove a revolutionary model and the impossibility of scientific progress.[1]

Against this background of competing positions, I shall summarize my own point of view. The history of science is a vigorous and rapidly rising discipline, but unlike the exact sciences it does not have a well-developed theory. I am not at all convinced that any of the theories of history can be applied to the history of science. From the perspective of science itself, the history of science is a young field of scholarship, still in a pre-Newtonian, largely inductive stage. But that does not have to be a handicap for understanding the history of science in a profound way. To that end, we need to do two things: (a) to "understand" thoroughly many of the individual main *events* in the history of a science (events, not speculative constructs such as the "Newtonian Revolution" or the "synthesis"); and (b) to see *connections* between many of these main events.

What is meant by "event"? Examples of events in the history of science would include the preparation or publication of a scientific paper, or the delivery of an influential speech (such as Niels Bohr's first pronouncement on complementarity at Como in 1927) and transcripts of discussions in the heat of battle, or the documented assurance of a specific discovery (such as the discovery of artificial radioactivity with slow neutrons on a certain day in October 1934 in Fermi's laboratory), or the preparation of an instrument, a letter, or a laboratory photograph (such as C. T. R. Wilson's in 1912), or a page of theoretical manuscript of a laboratory notebook (e.g., Millikan's), or the recording of an oral history interview conducted by a well-prepared historian of science. Each artifact is rooted in the act of doing science and is the preferably unselfconscious physical residue of an action at a given time; it is something that can be studied at a later time and may lend itself to a consensus among competent observers and interpreters who come to the case from different directions. An event in the history of science is in this sense analogous to what an elementary particle physicist calls an event, i.e., a trace on a photograph taken of a bubble chamber or spark chamber. The term is also related to the insistent use of the word *event* in Einstein's relativity paper of 1905; mentioning the word eleven times in a few paragraphs, Einstein implies that science should concern itself in the first place not with fundamentally metaphysically laden notions such as "matter" or "forces" but with what are now called intersections of world lines in space time.

Focusing on the events helps one avoid the mistake of attempting to impose umbrella conceptions on the raw material, far removed from the underlying factual base in the present state of the field. I would rather try to enter into the belief system inductively, from the sharable evidence of events that involved the personal struggle of ideas.

Different scholars will have different preferences for events that interest and excite them. For example, some opt for paradoxes or incongruities surrounding the event. Others may look for evidence of the operation of institutions, or whole professions, or social movements. Others still, scared of being charged with "psychologizing," make do only with the final printed product of a scientist and avoid documentary archives of its context or motivation. My own preference has often been to start with an artifactual residue of an individual's speculative attempts to understand a problem during the nascent phase, which Einstein called "the personal struggle." That is the period *before* the new work has come to some fruition, has been rationalized, pruned of all personal content, published, and absorbed in the mainstream of science through the mechanism of justification. This is, however, *after* the earliest and rarely documentable stirrings of the scientist's new idea. Insofar as I can find the necessary evidence, I tend to focus as much on the state of disorder the precedes the neat conclusion as on the conclusion itself. To study and describe the mind of the scientist in its nascent phase, I look for events that might be observed, as it were, through a keyhole in the door to the laboratory. In this spirit, Peter Medawar, in *The Art of the Soluble*, asked: "What sort of person is a scientist, and what kind of act of reasoning leads to scientific discovery and the enlargement of the understanding?"[2] He found the usual approaches too limited and issued a challenge: "What scientists *do* has never been the subject of scientific, that is, an ethological inquiry....It is no use looking to scientific 'papers,' for they not merely conceal but actively misrepresent the reasoning that goes into the work they describe....Only un-studied evidence will do—and that means listening at a keyhole."[3]

Often the event that attracts my attention is the writing of a private let-ter that was not meant to be widely shared, for example, young Albert Einstein's letter of March 1905 to his friend Conrad Habicht. In it, Ein-stein says he is working on several papers at the same time. The first pa-per, he says, concerns light radiation, and he calls it "very revolutionary" (one of the very few times Einstein used the word, and he is using it in the sense that in his opinion that work is not based on sound principles but only on a "heuristic" point of view). His second paper, he says, is on the size of the atom, determined from the diffusion phenomena. The third, he explains, is what we now remember as the paper on Brownian movement. And then he adds: "The fourth work is still in draft, and is an electrody-namics of moving bodies, using a modification of the teachings on space and time." (Note the word *modification*—announcing a work considered to be within the existing tradition, the very opposite of a revolutionary act.)

This letter became important because it opened one's eyes to the possi-bility of a connection among those four papers, each of which in its pub-lished form used to be seen as separate and of very different content. It sug-gested that one must reread those four publications and find the relationships among those seemingly isolated events. I shall come back to this point later.

Now imagine that a particular letter or article or other artifact, indicat-ing a puzzle or an entry into the nascent phase, has captured your atten-tion and has become your chosen "event." You must next proceed to give the proper *account* (in the sense of a descriptive portrait) of the event. To give the account of an event fully means, in my terminology, to begin to understand it. "Accounting" does not consist merely in providing a repli-cation of the letter, article, etc. Rather, it consists of finding and providing as many as possible of the separate *main components* that produced the chosen event. (The list of such components, as you will see, is a metaphor not so very different from the list of active forces present in the creation

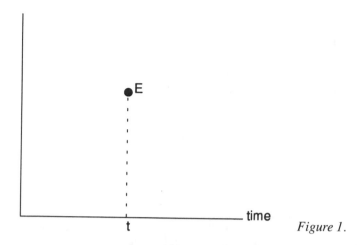

Figure 1.

of any work of scholarship, of literature, or of art.) Purely as a mnemonic device, let me represent the event E under study as a point in a plane, within orthogonal coordinates, the horizontal of which indicates time. E takes place at a given time t (or within an interval Δt) (Figure 1).

There are nine components which in principle make up the full account of any event E in the history of a science. It is unlikely that all nine can be described at once or by the same person engaged in the study of this case; but if an event becomes more fully understood by the profession as a whole, it is because more of the components have been brought in to contribute to the full account.

1. *First we must establish an inventory of the contemporaneous state of public, shared scientific knowledge (and ignorance) of the subject at time t.* What were the published "facts," data, theories, instrumental techniques, and widely believed laws and lore, whether or not they turned out later to have been erroneous? What did a scientist, or a group of them, believe to be the issue to be faced at a given time, the questions, the tools, indeed the scientific challenges within which this documentary contribution was made? We can call it establishing the historic, publicly available state of science at t (as seen by the scientist in question at that time t).

If we do not do this first, we shall find it almost impossible to avoid the trap of ahistoricity. As a warning, consider the remark of a physicist, writing in a review in the journal *Nature*, that it is an "old myth" that Boltzmann's constant and Avogadro's number were not reliably known before 1905. After all, the author says, Planck's published values in 1900 for these constants, based on his theory of black-body radiation, differed by only 3 percent from today's value. But while this statement is true nu-

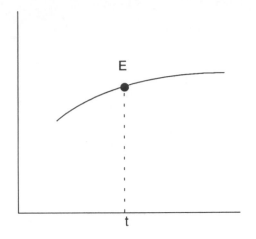

Figure 2.

merically, it is ahistorical, because in 1905, Planck's theory of radiation of 1900 was still widely ignored. A look at the handbooks of that day shows that these continued to print crude determinations of Avogadro's number while neglecting Planck's values.

2. *Next we need to establish the conceptual development of the public knowledge of a scientific branch, the time trajectory of the state of public ("shared") scientific knowledge leading up to, and if possible going well beyond, the time t chosen.* The event E is now, as it were, seen to be a point on this line, in a plane in which the horizontal dimension signifies time and the vertical dimension, only quite qualitatively, indicates increasingly understanding within a field of science (Figure 2).

This line is merely symbolic of the need to establish the conceptual antecedents. It is what Hans Reichenbach called the "context of justification" within which anything new will have to struggle for its place. Under this heading we would define the tradition or the area of controversy within which a scientist worked and expected his or her students to work. And we should also be able to trace the line beyond t, to get closer to the current scientific interpretation of the encounter E at t. This tracing of "conceptual development" is one of the most frequent activities for historians of science and for historically inclined science educators. This activity has the largest body of practitioners—it is relatively easy to do—and yet it harbors many dangers. Consider the example where the event E stands for the publication of Einstein's relativity paper of 1905.[4] Now the line going through E will have points that can be associated with the work of scientists before 1905, and for some years after 1905. Thus we would label one point on the line, corresponding to 1873, as the publication of

Maxwell's treatise. Higher on the curve and later in time would be points associated with Heinrich Hertz's experimental confirmation of Maxwell's electromagnetic theory, Michelson's experiment of 1881, Michelson and Morley's of 1887, and the publications of H. A. Lorentz and Henri Poincaré. And on points beyond 1905, we would place the publication of Walter Kaufmann of 1906, Max Planck's discussion of relativity in 1906, Minkowski's work of 1907, and so on.

This is how textbooks and some history of science courses deal with the development of relativity theory. But one must remember that this scheme is an artificial construction, one that imposes our current view of the "public" science during the period traced by the line. It has many flaws: First, it does not usually go back far enough—to be fair to the case, it should at least reach back to encompass the Galilean-Newtonian view on relativity, as well as Oersted's program of removing the didactic barriers between fields. Second, the usual tracing is in any case only a pedagogic, rationalized device, implying a linear, smoothed-out view of history, with one point or event on this curve causally and necessarily influencing the next one in time. In actual historic fact, nothing of this sort may have occurred. Rather, the interesting question may sometimes be about the actual discontinuities between parts of the curve rather than the continuities from one point to the next.

But if one bewares the trap of believing such fallacies, a great usefulness emerges from this treatment: One is led to ask what an innovator in fact knew of the related, publicly available developments prior to time t.

3. With this we come to the third component or aspect of an event. What we have done above for the state of public scientific knowledge, we must now do for *the "private" state of scientific knowledge of an individual for whom the event E arises at t*. That is to say, we must now study the less institutional, more ephemeral personal aspects of the scientific activity E at t. We are now looking, to the extent possible, at all material from the workshop and in the desk of a scientist at period t—the object that caught your attention in the first place, the surviving letters, drafts, personal laboratory notes, abandoned equipment, etc. For the sciences more than most other fields, these links to the past are fragile, making it difficult in many cases to provide the "context of discovery." In many of our universities, where every scrap of paper from literary scholars seems welcome in the archives, it has been more difficult to find a home for scientific correspondence, laboratory records, and apparatus.

Moreover, the need for documentation is not necessarily appreciated or understood by the scientists themselves. Some are still at best impatient with such studies (as with the whole study of what they regard as the "merely personal"). There are evident sociological reasons for that neglect and impatience. The very institutions of science, the selection and training of young scientists, and the internalized image of science are all designed to minimize attention to the personal scientific activity prior to publication. Indeed, the success of science as a sharable activity is connected with the conscious down-playing of the private struggle.

An example of a major scientist reconsidering the value of the past—now happily getting more common—is in the introductory paragraph of Richard P. Feynman's lecture on December 11, 1965, after having been awarded the Nobel Prize the previous day:[5]

> We have a habit in writing articles published in scientific journals to make the work as finished as possible, to cover up all the tracks, to not worry about the blind alleys or to describe how you had the wrong idea first, and so on. So there isn't any place to publish, in a dignified manner, what you actually did in order to get to do the work, although there has been in these days some interest in this kind of thing. Since winning the prize is a personal thing, I thought I could be excused in this particular situation, if I were to talk personally about my relationship to quantum electrodynamics, rather than to discuss the subject itself in a refined and finished fashion. Furthermore, since there are three people who have won the prize in physics, if they are all going to be talking about quantum electrodynamics itself, one might become bored with the subject. So, what I would like to tell you about today is the sequence of events, really the sequence of ideas, which occurred, and by which I finally came out the other end with an unsolved problem for which I ultimately received a prize.

Another obstacle to ready access to documents that might reveal the "private" state of scientific knowledge, particularly in its nascent phase, is the apparent contradiction between the seemingly illogical nature of actual personal discovery and the logical nature of well-developed scientific concepts. This contradiction is perceived by some scientists and philosophers as a threat to the very foundations of science and to rationality itself. The vogue of some philosophers of science who attempted a "rational reconstruction" of actual scientific work seemed to have been partly motivated in this manner.

The alternative path is not easy either. In one of his interviews, Einstein urged historians of science to concentrate on comprehending what scientists were aiming at, "how they thought and wrestled with their problems." But he pointed out two difficulties. One was that the scholar would have to have sufficient insight and a kind of empathy (he called it a "*Fingerspitzengefühl*"), both for the content of science and for the process of scientific research, the more so as solid facts about the creative phase are likely to be few. Second, Einstein advised that, as in physics itself, the solution to historical problems may have to come by very "indirect means," and the best outcome to be hoped for is not certainty but only a good probability of being "very likely correct anyway." Historians have come to see the arrogance of claiming that they can reconstruct a historic event "as it really happened." We cannot attain absolute certainty, either in science or in the history of science. But in both cases one may succeed by taking informed, reasonable risks.

4. *Again, as with public science we must next establish the time trajectory of the personal scientific activity under study.* Here we look for the interests and accomplishments and failures of the persons involved over a large portion of their careers. It includes both the period of preparation, on one side of t, and the period of harvesting, on the other side. The work of an individual at t is seen with greater understanding if we have followed the development of his or her personal style prior to t. This new trajectory symbolizes the development of the private or personal science of the person responsible for the event E. Turning again to Einstein, we would identify a point for the year 1894–95, when he wrote his first scientific essay (on the ether, based on reading Heinrich Hertz). Next we would follow his thought experiment at the age of about 16 at school in Aarau 1895–96, then his reading of Ernst Mach and Hume and others, starting around 1897. Next, perhaps we would consider his confession of "despair" of trying to do fundamental physics from the experiential base upward, in 1900; and then the work leading to the publication of his various great papers in 1905 .

Now we see an event E as the intersection of *two trajectories*, one for what we called the private science, a line labeled S_1, and one for the public science, to which trajectory we might give the symbol S_2 (Figure 3).

This point of view opens up immediately some new problems worth working on. One of these would be the acknowledged intersection of S_1 and S_2, with the result that important new names would have to be added to the S_2 trajectory, such as the influence of August Föppl, who has usually been forgotten in the accounts of "conceptual development" along S_2, but whose book greatly influenced Einstein as a young man. Another problem to consider would be the *lack* of much influence of a work that, while prominent on the profession's S_2 line, happened not to have been absorbed into the individual's S_1 line. This would apply to Lorentz's 1904 formulation of the theory of the electron, which Einstein hadn't read, and the Michelson-Morley experiment, for which Einstein repeatedly, and consistently over many years, specified that it had at most an indirect, and certainly not a crucial, influence on him.

Now let me give an example of the importance of tracing the S_1 trajec-

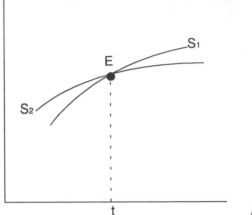

Figure 3.

tory. When you read Einstein's early publications and the letters in the archives, you see that the trajectory that brought him to the relativity theory in the miraculous spring of 1905 began in an apparently unpromising way. Einstein's first published article (1901) was entitled "Consequences of the Capillarity Phenomena." For a very young man in need of a job and a career, this was a curious choice. At the time, all the excitement in physics lay in a quite different direction. It was just a few years after the discovery of X-rays, radioactivity, and the electron. New experimental findings and new theories chased one another at a dizzy pace. In comparison, capillarity was an old and rather boring subject. But if we study this paper more carefully, we discover something important. The problem to which Einstein is attending, in this first paper and in the next one, is "the problem of molecular forces." He starts with the promise to proceed from the simplest assumptions about the nature of molecular forces of attraction. In this work, he says, "I shall let myself be guided by analogy with gravitational forces." What interests him, as he writes in a letter to his friend Marcel Grossmann (April 14, 1901), cited also in the previous chapter, is "the question concerning the relationship of molecular forces with Newtonian forces at a distance." Now, that shows great ambition! Einstein's lifelong interest is making its first appearance here: the program of the unification of the various forces of nature. He felt he was working on an important problem. When he writes that "It is a magnificent feeling to recognize the unity of a complex of phenomena that to direct observation appear to be quite separate things," Einstein is but 22 years old, but it is already a familiar Einstein, here searching for bridges between the phenomena of microphysics and macrophysics.

Now let us look again at those major papers of 1905, which Einstein sent off to the *Annalen der Physik* at intervals of less than eight weeks. When I first became interested in this case, it struck me that while the papers—on the quantum theory of light, on Brownian movement, and on relativity theory—seemed to be in quite different fields, the letter to Conrad Habicht I referred to earlier shows that it may have been no accident that they were all done at about the same time. In fact, all can be traced in good part to the same general problem, namely, fluctuation phenomena. Indeed, in the archives I found a letter Einstein wrote to Max von Laue in 1952 in which the connection is indicated. Einstein discusses there the new edition of von Laue's old textbook on relativity theory and registers an objection:

> When one goes through your account of verifications of the special relativity theory, one gets the impression Maxwell's theory was unchallengeable. But already in 1905 I knew with certainty that Maxwell's theory leads to wrong fluctuations in radiation pressure, and hence to an incorrect Brownian movement of a mirror [suspended in] a Planckian radiation cavity. In my opinion, one cannot get around ascribing to radiation an objective atomistic structure, which of course does not fit into the framework of Maxwell's theory.

Here we see explicitly the chief connection between Einstein's work on Brownian motion of suspended particles, the quantum structure of radiation, and his more general reconsideration of "the electromagnetic foundations of physics" itself. In short, to understand the relativity event E, we need to understand it as part of a more extensive research program.

Moreover, the style of the separate papers, despite their seemingly diverse topics, is essentially the same. Contrary to the sequence one finds in many of the best essays of that time, for example, in H. A. Lorentz's publications on electromagnetic phenomena, Einstein does not start with a review of a puzzle posed by some new and difficult-to-understand experimental fact. Rather, he begins in each case with a statement of his dissatisfaction with what he perceives to be formal asymmetries or other incongruities (which, to other eyes, must have seemed to be predominantly of an aesthetic nature rather than of scientific importance). He then proposes a principle of great generality—one that applies to apparently different parts of physics. Third, he shows that this principle helps remove, as one of the deduced consequences, his initial dissatisfaction. And fourth, at the end of each paper he proposes a small number of predictions that should be experimentally verifiable.

To review: we have traced two lines, two trajectories, one for public science and one for private science; and the event that interests us at a given time begins to be understood as an *occurrence at the intersection of these two trajectories.* We now need to indicate at least briefly the tracing of other strands, the first of which is intertwined with S_1.

5. Here I refer to the *biographical (nonscientific) development* of the scientist under study. The special case of psychobiography is a new and uncertain field, but we have begun to get good material, such as Frank Manuel's and R. S. Westfall's on Isaac Newton, and Erik Erikson's on young Einstein. I shall give an example, in a later chapter in this volume, that illuminates the events at two points in the career of J. Robert Oppenheimer by relating them to his prior biographical experiences.

6. Another component of a scientific event is its *social setting*, for example, the effect of the educational system on the preparation of scientists in different countries. A case in point is the reception given to the relativity theory in its early years in Germany, Britain, France, and the United States. The differences were enormous during the first decade or two, and a large part of the explanation is found in the differences of the educational systems. For example, in Britain, the preparation of physicists concentrated heavily on the properties of the ether, making it difficult for many of them to break away from that concept. In France, the hierarchical and pyramidal structure of the profession was such that the opposition of Henri Poincaré, the premier scientist in his time, to Einstein's relativity made it very unadvisable for anyone to publish in this field until after Poincaré's death in 1912. In Germany, the fierce competition between autonomous universities (rather than, as in France, the primacy of one university in the country) assured a variety of lively responses. And in the United States, the pragmatic style of understanding the science permitted a surprisingly early acceptance of relativity (though many were under the impression that it was fully supported from the beginning by an experimental base).

Under the same heading of sociological setting, we should also look into the role played in a scientific advance by such effects as that of teamwork in science; the link between academic scientific research and research of interest to industry and the military; and the system for funding research. It would be difficult, for example, to understand the burgeoning of physics research in the United States in the 1920s without paying attention to the role which the grants and policies of private philanthropic foundations such as the Rockefeller and the Guggenheim played at that time. Foundations financed a large fraction of American (and indeed foreign) physicists studying and doing research during the explosive growth of quantum physics. By doing so they helped to create a "critical mass" of scientists whose choice and pace of research were quite different from what they would have been without their grants.

7. A seventh strand deals with the *cultural developments* outside science and the *ideological currents or political events* that influence the work of scientists. Obvious cases spring to mind: the link such scientists as Galileo, Kepler, and Newton saw between science and theology; the link between politics and science in such cases as that of Lysenko, or of the German physicists under the influence of fascism from the 1920s on. But the point is more general, and sometimes the subject of continuing debate. One of these is the case of quantum mechanics; Paul Forman started a discussion of the degree to which developments in the German intellectual environment, including the vogue then enjoyed by Oswald Spengler's book *The Decline of the West* and the rise of revolutionary ideologies, prepared some German scientists to abandon the classical principle of causality in the 1920s.

Both this strand and the previously noted influence of the social setting can be so attractive to some students of the history of science that they might be tempted to see all cases as the result of those components alone—an extreme form of the "constructivist" approach. This is as limited a view as its polar opposite—which is narrowly "internalist," focused only on the published product of a scientist's labors.

8. In certain cases the work under study may well be illuminated by an analysis of the philosophical component, particularly of the *epistemological suppositions and the logical structure*. The philosophical world view of a scientist surely is as important as, for example, his understanding of the mathematical tools of his trade. I am thinking here of how Einstein confessed to have been influenced by his reading of Ernst Mach and David Hume, or of R. A. Millikan's robust belief in the reality of atoms despite the anti-atomistic teaching of his doctorate supervisor, Michael Pupin, and of others at the time. Or consider that in Germany, and practically nowhere else, the early discussions of quantum mechanics were heavily penetrated by quasi-metaphysical debates on *Anschaulichkeit*—questions on how far scientific concepts can depart from visualizable institutions—which stemmed chiefly from the hold on German intellectuals of ideas derived from their early reading of Immanuel Kant.

9. So far, we have a list of eight components that in principle can be discerned in the activities from which a given event arises. Last but not least, I turn to the ninth tool for the analysis of a scientific work, which I have termed *thematic analysis*.[6] By this I refer to the often unconfessed or unconscious guiding presupposition a scientist adopts *without* being forced to do so by either data or current theory. An example is Heisenberg's decision to establish a physics built on the thema of discontinuity—thereby freeing physics from what he called, in a letter to Pauli of January 1925, the "swindle" of working with a mixture of quantum rules and clas-

sical physics, as Bohr and Sommerfeld still preferred. This decision of Heisenberg led in turn to his abandoning the customary visualization by which space and matter were thought to be in principle continuous.

Another, and opposite, example of the role of a thematic presupposition is found in Einstein's outburst to Max Born (December 3, 1947). Despite the great initial success of quantum theory, Einstein was obstinately unconverted as regards the fundamentality of probability. And so he wrote:

> I cannot substantiate my attitude of physics in such a manner that you would find it in any way rational. I see of course that the statistical inter-pretation...has a considerable content of truth, [but] I am absolutely con-vinced that one will eventually arrive at a theory in which the objects con-nected by law are not probabilities but facts processed by thought, as one took for granted only a short time ago. However, I cannot provide logical argu-ments for my conviction, but can only call on my little finger as a witness, which cannot claim any authority to be respected outside my own skin.[7]

I have listed earlier the presuppositions that guided Einstein throughout his life: the primacy of formal (rather than materialistic) explanation; unity or unification on a cosmological scale; logical parsimony and neces-sity; symmetry; simplicity; causality; completeness; continuum rather than atomistic discreteness; and, of course, constancy and invariance. These are the types of themata that may explain in specific cases why a scientist will obstinately continue his or her work in a given direction even in the face of ambiguous or contradictory evidence. As noted, they explain why Einstein refused to accept theories that were well supported by correlation with phenomena but were based on Bohr's opposing the-matic presuppositions.

As Einstein's reluctance to accept probability shows, thematic presup-positions, in themselves not verifiable or falsifiable, can lead one astray if they are held too long against mounting evidence against them. But at least in the nascent phase of one's work, thematic presuppositions—which Einstein referred to as "categories or schemes of thought"—are necessary for most scientists. They are in the logical sense "free conven-tions," hence not unalterable or conditioned by the very nature of our un-derstanding (contrary to Immanuel Kant's categories).

The realization of the thematic origins of scientific thought corrected an appealing but simplistic notion about scientific method that was cur-rent in earlier times, and still infects some pedagogic presentations—the notion that the individual scientist always must, and can, start out utterly free from all preconceptions, somewhat as Thomas H. Huxley still thought possible:

Science seems to me to teach in the highest and strongest manner the great truth which is embodied in the Christian conception of entire surrender to the will of God. Sit down before fact as a little child, be prepared to give up every preconceived notion, follow humbly wherever and to whatever abysses nature leads, or you shall learn nothing. I have only begun to learn content and peace of mind since I have resolved at all risks to do this.[8]

What does save science from falling victim to inappropriate presuppositions are of course the chastening roles both of the coordination with experiment and of the multiple cross-check of any finding by other scientists who may have started with quite different presuppositions. That is why it is now generally thought that Bohr, contrary to Einstein, was right in accepting the thema of fundamental probabilism at the base of atomic theory. Hence, as Dirac pungently put it,[9] "At the present time, one must say that, according to Heisenberg's quantum mechanics, we must accept the Bohr interpretation. Any student who is working for an exam must adopt this interpretation if he is to be successful in his exams." But then Dirac added a warning that one cannot predict which of two antithetical themata will win out in the long run:

Once he has passed his exams, he may think more freely about it, and then he may be inclined to feel the force of Einstein's argument....It seems clear that the present quantum mechanics is not in its final form....Some day a new quantum mechanics, a relativistic one, will be discovered, in which we will not have these infinities occurring at all. It might very well be that the new quantum mechanics will have determinism in the way that Einstein wanted....even though for the time being physicists have to accept the Bohr probability interpretation, especially if they have examinations in front of them.

To recapitulate my view how the scientific mind works to produce an event in the history of science, it is not sufficient to deal only with the resolution of the finding under examination along the two axes of empirical or phenomenic content and analytical content—in shorthand, meter readings and mathematical tautologies, respectively. These define, as it were, the two axes, x and y, in the plane of usual discourse about scientific work. But there is also a crucial z-axis, perpendicular to the plane, along which the thematic presuppositions are resolvable in the specific case, and I have found they can play a dominant role in the initiation and acceptance or rejection of key scientific insights. A thematic attachment may favor one or the other of two (or sometimes three) antithetical themata (i.e., θ or $\bar{\theta}$), and these are often traceable to very early speculations, such as those concerning constancy versus change in the traditions of Parmenides and Heraclitus. The historian of science trying to understand a specific case or event should therefore be alert to possible thematic choices made by the scientist: choices between experience and symbol formation, complexity and simplicity, reductionism and holism, discontinuity and continuum, hierarchical levels and unity, evolution/devolution/steady state, mechanistic/materialistic/mathematical models, causality/probabilism, analysis/synthesis, and so on.

At first level, the full understanding of a significant event in the history of science will thus be accomplished by description and analysis of the event in terms of these nine components. Since scholarship in the history of science is a communal, collaborative effort, as is science itself, each of us may be privileged to contribute only a few of these components to a few of the significant events. But in time, the depth of understanding grows by the overlap of these individual contributions. When a historian of science has achieved at least a partial "understanding" of many events $(E_1, E_2, E_3...)$ scattered through time, it becomes possible for that person to achieve an understanding of the field at a second level, in a more profound way. *Connections* between individual events begin to appear, analogous to understanding connecting routes found on a map. One may learn, for example, that both Einstein and Millikan were influenced, though in quite different ways, by Ernst Mach. Ernest Rutherford and Enrico Fermi, working in different countries on rather different problems, nevertheless influenced each other's work at a crucial period. And of course an event E_1 in the life of a given scientist (say, Bohr's papers of 1913) can have profound and complex relationships with the event E_2 in the same scientist's later work (e.g., Bohr's announcement of complementarity in 1927).

From this point of view, the growth of the field of scholarship as a whole is governed by the accumulation of the deeper understanding of a great and increasing number of separate-seeming events and subsequently of the connections between them. What to the novice may well have seemed a chaotic, infinite, and almost arbitrary collection of publications, controversies, letters, devices, techniques, etc., is thus transformed into the perception of coherent patterns that help explain the development of a scientific field. It is through this perception that one may say one is beginning to understand the history of a science.

References

1. The *Guide to the History of Science*, published by the History of Science Society, lists the following subsets of subject interests of its members: Bibliography and Archives, Historio-

graphy, Biography, Philosophy of Science, Scientific Institutions, Instruments, Social Relations of Science, Humanistic Relations of Science, History of Science in Education, Science Policy Studies, Mathematics, Astronomy, Physics, Chemistry, Earth Sciences, Exploration and Expeditions, Biological Sciences, Social Sciences, Medicine, Pharmacy, Technology and Engineering, Astrology, Alchemy, Music, Nonindustrial Cultures, Ancient Near East, Greek and Roman Antiquity, Medieval Europe, Islam, Middle and Near East to 1600, American Science, India, East Asia, sixteenth-seventeenth century, eighteenth century, nineteenth-early twentieth century, Contemporary Science, Technology, and Medicine.

2. P. W. Medawar, *The Art of the Soluble* (London: Methuen & Co., Ltd., 1967), p. 7.
3. Ibid, pp. 151, 155.
4. In this section, I refer to certain details in case studies on relativity theory to be found in three of my books: *Thematic Origins of Scientific Thought: Kepler to Einstein* (Harvard University Press, 1988), *The Scientific Imagination: Case Studies* (Cambridge University Press, 1978), and *The Advancement of Science, and Its Burdens: The Jefferson Lecture and Other Essays* (Cambridge University Press, 1986).
5. Richard P. Feynman, "The Development of the Space-Time View of Quantum Electrodynamics," in *Les Prix Nobel en 1965* (Stockholm: Imprimerie Royal P. A. Norstedt & Sönner, 1966).
6. A detailed discussion of this conception, and studies illustrating its influence in specific cases, are given in the three books mentioned in note 4, above.
7. Einstein to Born, March 3, 1947. Translated from Albert Einstein und Hedwig und Max Born, *Briefwechsel 1916–1955* (München: Nymphenburger Verlagshandlung, 1969), p. 215. See also *The Born-Einstein Letters*, trans. Irene Born (New York: Walker and Co., 1971), p. 158.
8. Quoted in Leonard Huxley, *The Life and Letters of Thomas Henry Huxley* (London: MacMillan, 1900).
9. P.A.M. Dirac, "The Early Years of Relativity," in G. Holton and Y. Elkana (eds.), *Albert Einstein, Historical and Cultural Perspectives* (Princeton, NJ: Princeton University Press, 1982), pp. 84–85.

Part Three:
PERSONALITIES
IN PHYSICS

The Trials of
J. Robert Oppenheimer

Throughout his career Robert Oppenheimer (1904–1967) managed to be at the center of excitement. History remembers him most for his leadership in 1943–45, as director of the project that propelled the world into the nuclear age; indeed, he was the first person to push the nuclear button, on 16 July 1945, at the Alamagordo test. But before that he was also one of the crucial figures in bringing the United States into the front rank of physical science. For a time, he served as one of the foremost academic "statesmen of science" in the change from disengaged science to science as the helpmate of national and international policy. And he played a role in other episodes of twentieth-century American history, such as the rise and fall of McCarthyism and the acceptance of Jews into the academic professoriate.

All this, and his own excellent work in the physics of quantum field theory, cosmic rays, and nuclear phenomena, should be enough to assure Oppenheimer a place in history. But there is another reason for the recurrent fascination with him. He was often a spellbinding presence; his intelligence and his complex personality shone through hypnotic eyes. It is generally agreed that no one else could have directed so well the large group of prima donna scientists assembled at Los Alamos under the difficult and panic-evoking conditions of war. Even now there is a quasi-charismatic air about his memory, exciting vociferous loyalty among friends and disciples and strong condemnation from opponents. On the popular level, he is seen as a tragic hero who, chastened by the discovery of "sin" in the light of the A-bomb, wrestled mightily with his conscience but in the end made yet another Faustian bargain when Edward Teller suggested a "sweet" solution to the problem of making an H-bomb. Oppenheimer provides an early example of the contemporary dilemma: when society tries to harness the drives of science and technology toward omniscience and omnipotence, it often hitches them to short-range purposes that have quite unforeseeable long-range consequences.

Young Man Oppenheimer

On each of these topics, an elegant book assembled by Alice K. Smith and Charles Weiner[1] provides some illumination. It gives us over one hundred letters from newly found caches of Oppenheimer's correspondence, starting when Oppenheimer was but eighteen years old and ending twenty-three years later, with the previously unpublished text of his farewell address to the Los Alamos scientists in November 1945. The editors supplemented the letters with extensive research and commentary based on earlier interviews with Oppenheimer and on thirty-three interviews with his contemporaries. We shall find all this a good beginning for an attempt to understand Oppenheimer.

Our attention is drawn first to the early period, partly because we might find precursors of his later development and partly because, from the beginning, Oppenheimer appears to be an uncommon person, so different from the rather dull-seeming lot that haunts C. P. Snow's novels and, in fact, most laboratory corridors. The correspondence starts when the awkward eighteen-year-old, on entering Harvard College, immediately writes to his beloved former teacher at the high school of the Ethical Culture Society in New York, Herbert W. Smith, who continued to play the role of mentor, companion, and father-confessor for many years to come. Oppenheimer's father was a well-to-do businessman and his mother a trained painter. Robert had a *jeunesse dorée* of the best sort, his day starting with breakfast under the Van Goghs in a Riverside Drive apartment.

At school, learning everything easily, writing stories and poetry, Oppenheimer developed a baroque and exaggerated style, evident in many of the early letters, which may have later penetrated his physics—he confessed that "I probably still had a fascination with formalism and complication" before he fully realized the need for "simplicity and clarity." His ambition at college was, first, to study chemistry or mineralogy "with the idea of becoming a mineral engineer." Apparently he had become fascinated with science as a child when, on a visit to his grandfather in Germany, he was given a collection of minerals. In his family one did things in style; so as a high school graduation gift the boy was staked to a prospecting trip to Czechoslovakia—to the mining center in Joachimsthal, famous for its pitchblende from which uranium salt had long been extracted for the manufacture of glass. It was from there that tons of pitchblende residue had been obtained by Pierre and Marie Curie, who then discovered in it the mysterious substance they called radium. The same area was to be a main supply of uranium, many years later, for the German attempt to make an atomic weapon. Indeed, the German embargo on the export of uranium ore from Joachimsthal was one of the telltale signs noted by Einstein in his letter of 2 August 1939 to Roosevelt, warning that the Germans were on their way.

Oppenheimer visited that mine at a time almost precisely halfway between the Curies' labors and Einstein's letter—one of the mysterious symmetries in his life trajectory, a pattern that normally we associate only with staged plays. On that excursion, he contracted a nearly fatal case of dysentery, which exacerbated a predisposition to bouts of colitis, melancholia, and periods of deep depression. Again, a strange irony: Having been forced to spend a year at home to recuperate and thereby miss a year before entering college, young Robert was sent on a refreshing trip to New Mexico and the West, Herbert Smith serving as companion. The boy fell in love with the wild Southwest—and incidentally came upon a beautiful and isolated mesa on which was located the Los Alamos Ranch School. Twenty years later, Oppenheimer took General Leslie Groves to that spot one November afternoon, to propose it as the site for the secret laboratory.

Once in college at Harvard, Robert immersed himself in chemistry, mathematics, and physics. But it was not clear which of these would win out, and he also stretched in many other directions, continuing his study of languages (he knew Latin, Greek, French, Spanish, and German, and later added Dutch and serious study of Chinese and Sanskrit) and discovering the Widener Library, literature, A. N. Whitehead's philosophy seminar, excursions to Cape Ann. To Smith he wrote, "I labor and write innumerable theses, notes, poems, stories, and junk; I go to the math lib. and read and to the phil. lib. and divide my time between Meinherr Russell and the contemplation of a most beautiful and lovely lady who is writing a thesis on Spinoza—charmingly ironic at that, don't you think? I make stenches in three different labs, listen to Allart gossip about Racine, serve tea and talk learnedly to a few lost souls, go off for the weekend to distill the low-grade energy into laughter and exhaustion, read Greek, commit faux pas, search my desk for letters, and wish I were dead."

The various fragments of Oppenheimer's soul were evidently waiting for a magnetic field to line them up. The magnet was his encounter with the experimental physicist P. W. Bridgman, who was then also developing his operationalist philosophy of science. Later Oppenheimer recalled that with regard to science, Bridgman's course represented "the great point of my time" at college, for Bridgman "didn't articulate a philosophic point of view, but he lived it…. He was a man to whom one wanted to be an apprentice."

First, however, Oppenheimer was heading for terrible disappointment, the first of his severe trials. Bridgman later told me with some amusement that young Oppenheimer had proposed working on an experimental problem in his laboratory, and at the beginning he "didn't know one end of the soldering iron from the other." The delicate suspensions of the galvanometers had to be replaced constantly when this apprentice was using them. He stuck with it, measuring the electric conductivity of alloys at high pressures—a kind of apotheosis of his initial interest in minerals and crystals.

But in the early 1920s, the lack of sufficient skill in the laboratory raised serious doubts about the ultimate promise of a would-be physicist in the United States. (It was still quite unusual for a young American to think of himself as a *theoretical* physicist.) Oppenheimer had his heart set on going to Ernest Rutherford's vigorous, renowned laboratory at Cambridge, England; but he was not accepted for work under Rutherford and had to settle for a place with the septuagenarian J. J. Thomson. That was a major blow, to which was added a bigger one, setting the stage for a devastating year. As Oppenheimer himself said in a letter of about that time, he now recognized that "my genre, whatever it is, is not experimental science." That realization may help explain in part the depression and identity confusion which overtook the young man at that point, just when he was engaged in "the difficult business of making myself for a career" in a place and at a time that accentuated for him his weaknesses and deficiencies. He still had not made the central discovery that every young person must make, in Max Weber's words, to "find and obey the demon that holds the fibers of one's very life."

At twenty-one, Oppenheimer was struggling with the anguishing problem of feeling a basic need to look for central scientific questions, but neither seeing them yet with clarity nor having the ability to cope with the disappointment of remaining on the periphery. A few years later, in one of his beautiful letters to his brother Frank, he hinted at the eventual internal resolution: "I know very well, surely, that physics has a beauty which no other science can match, a rigor, an austerity and depth"; its study induces, and demands, a kind of mental discipline that at its best helps one "achieve serenity, and a certain small but precious measure of freedom from which the accidents of incarnation, and charity, and that detachment which preserves the world which it renounces…. We come a little to see the world without the gross distortion of personal desire…."

Herbert Smith had warned an American friend in Cambridge that Robert, on leaving his earlier, pampered life at college might find "frigid England hellish, socially and climatically," and that if allowed to be over-awed, "he'd merely cease to think his own life worth living." That evidently diagnosed part of the problem, but not the fundamental part. Rather, it was first Robert's bad luck, and later his rescue, that this labile, vulnerable, and brilliant young person passed through a personal crisis in the mid-1920s at the very time when physics itself was going through its own great transformation, the discovery of quantum mechanics—a period which older scientists, from Niels Bohr to Werner Heisenberg, were calling frankly a time of "despair." The difficulty and the breakneck pace of the new science, together with the tantalizing new glimpses of the great work waiting to be done, clearly intensified in young man Oppenheimer what can euphemistically be called a serious problem of temperament; and at the same time, this period would put to a test what one of his friends termed Robert's "ability to bring himself to figure out what his trouble was and to deal with it."

Even at the best of times, becoming a competent scientist is a hard task—as Oppenheimer himself once counseled, it is "like climbing a mountain in a tunnel: you wouldn't know whether you were coming out above the valley or whether you were ever coming out at all." But as it happened, for Robert the turbulences of the history of physics and of his own psychohistory had come together in a most frustrating way and culminated, during a Christmas holiday trip to Paris, in an episode that showed the severity of the psychological crisis. Smith and Weiner write that, apparently without warning, "Robert suddenly leapt upon [Francis] Ferguson [a friend and fellow student from Harvard days] with the clear intention of strangling him.... The uncharacteristic display of violence, combined with Robert's despair over his inept performance in the laboratory, and confidences from unsatisfactory sexual ventures, convinced Francis that his friend was seriously troubled."

Subsequently, there were other melodramatic episodes, periodic visits to a Cambridge psychiatrist, and evidence of continued depression. That spring, Robert told his friend John Edsall that "he had *dementia praecox* and that his psychiatrist had dismissed him because in a case like this further analysis would do more harm than good." To another friend he said he was leaving his psychiatrist because the fellow was too stupid to follow him, and that he knew more about his own troubles than the doctor did.

A word needs to be said about the use in those days of the phrase *dementia praecox*. The same symptoms might now be called borderline schizophrenia, a term that lacks the stigma of deterioration, debility, and irreversibility. Through the work of Erik Erikson and others on the identity formation of young persons, we are now more ready to deal with such episodes. For cases in which, in the past, young people were judged to suffer from a "chronic malignant disturbance," we now suspect, in Eriksonian terms, a somewhat delayed developmental crisis. The young person is engaged in two tasks. One is the working through of a "negative identity"—the sum of the fragments which the individual has to submerge in himself as undesirable or irreconcilable; and the other is the development of a positive identity to preserve the promise of a traditionally assured wholeness. Erikson wrote specifically about the rage that can be aroused by threatened identity loss and that can explode in arbitrary destructiveness, as well as the more traditional explanation of dangerous or radical alternation between loving and hateful tendencies during transference.

Some day there may exist enough material and skill to attempt a psychobiographical study of Oppenheimer. But it is clear already from these letters that, once he had made the decision not to go back into a laboratory but to throw himself into the grand theoretical problems then attracting some of the best European physicists, Oppenheimer had essentially embarked on a course of self-therapy. Not infrequently, those of us who as teachers and mentors come to know young would-be scientists during their crucial formative years see some version of this process of self-identification taking place before our very eyes. For Oppenheimer, joining in the great intellectual drama of transforming physics helped resolve his dilemmas, once he achieved some success. Some psychological damage remained, however, not least a vulnerability that ran through his personality like a geological fault, to be revealed at another earthquake—as we shall see.

But the letters of the young Oppenheimer, and reminiscences recorded later both by Oppenheimer and his friends of that period, show that the years 1926–27, spent at the University of Göttingen, produced an inner alignment that formed the armature of his mature self. He met like-minded, challenging people, such as Richard Courant, Werner Heisenberg, Gregor Wentzel, and Wolfgang Pauli—a support group with which, he said, "I began to have some conversations." The formal and informal education, the development of his insight in frontier problems, now proceeded at an incredible pace. Early in 1927, he could write from Göttingen (perhaps with much satisfaction) a letter to his former mentor Bridgman, explaining why Bridgman's rather classical theory of metallic conduction—on the basis of which Oppenheimer had embarked on his ill-fated laboratory work—had to be discarded in order to take into account the new quantum mechanics. At the age of twenty-three he had essentially achieved his goal of making himself "for a career"—if not for a serene life.

Oppenheimer's main professor at Göttingen was, as it happened, Max Born—no easy taskmaster. At an early stage of his career, Enrico Fermi had been so intimidated by Born that Fermi had considered the idea of giving up physics, before being rescued by finding sympathetic tutelage under Paul Ehrenfest in Holland. Born had encountered Oppenheimer on a visit to the Cavendish Laboratory, and they discovered that they were both interested in the same problem, the light spectrum radiated from molecules. Within a few months of moving to Göttingen, Oppenheimer had finished his doctoral thesis; the resulting article on the quantum theory of continuous spectra was received at the *Zeitschrift für Physik* on Christmas Eve, 1926. Within a year, he had written four more articles, one of them, with Max Born, a quantum-theoretical treatment of the behavior of molecules that has remained in use decades later (the "Born-Oppenheimer Approximation"). Having passed victoriously through his ordeal, Oppenheimer returned to the United States as a kind of young guru of the new physics. From many job offers he chose a university at which no one was working in the new quantum mechanics, where he could begin his career of leadership of a major school of theoretical physics. But it is most significant that he also cultivated genuine working relationships with experimental physicists—the happy harvest of his earlier struggles in laboratories, and also a characteristic of the opportunities of the American university system, which did not separate the theoretical and experimental departments in the manner standard in European instruction and research.

As it turned out, Oppenheimer's range of interests was to be of great importance to his effectiveness as leader of the Los Alamos project; at that point, the very fate of the civilized world may have depended on it. For, precisely as Einstein and his colleagues had feared in August 1939, the Germans did start on a project to build a nuclear weapon; they began considerably before the antifascist alliance and had the advantage of excellent and experienced nuclear scientists, such as Otto Hahn, in whose laboratory nuclear fission had first been discovered. Obviously world history would have been very different if the Germans, working under the leadership of Heisenberg, had succeeded and the American-British-French effort had failed, or both had failed, or both had succeeded. Why, then, the actual outcome?

This question has been much analyzed, and many plausible reasons proposed, from the lack of sufficient support at the highest political level in Germany, to the Germans' failure to engage in the large industrial operation necessary for the actual production. Such factors undoubtedly played some role. But another factor—rarely noted yet made more evident by the story of Oppenheimer's career formation, as reflected in his letters—must be the difference in scientific styles that characterized the respective efforts, and in the sociology of the scientific community within which they worked.

In the German project, top responsibility and many of the essential ideas were those of Heisenberg, a theoretical physicist who, in the tradition of his country, had been able to avoid serious training in experimental physics during his formative years. The original design of the nuclear reactor pile, strongly influenced by Heisenberg's own ideas, turned out to be quite impractical. Also, at least one and probably two crucial experimental measurements at the outset of the project—the way neutrons diffuse in the moderator lattice of the pile, and the number of neutrons that are freed during fission—were botched; furthermore, the mistake was not effectively challenged, perhaps because in the hierarchical German system, transferred from academia to the weapons project, challenges from knowledgeable younger people in the laboratory were quite difficult to initiate.

On the other side, one of Oppenheimer's first acts on assuming leadership of the Los Alamos group—in fact a condition he made on accepting the assignment—was that an experimental physicist (John H. Manley) should be provided to assist him. In addition, Oppenheimer sought out skilled experimentalists of any age or rank. One of the earliest tasks the group tackled was to relate the theoretical calculations of nuclear reactions to the experimental data, thereby allowing an estimate of the critical mass of the fission weapon and its efficiency. The interlacing of the theoretical and experimental aspects was complete under Oppenheimer's influence, and natural for all who worked with him. Moreover, Oppenheimer, the experimentalist *manqué,* was able to keep informed about both theory and experiment, to stretch and supplement his expertise from nuclear physics to metallurgy and ballistics, and by assignment and example he helped others to do the same. His taste and skills across this whole spectrum provided much of the glue that held the enterprise together. Some roots of this success—for better or worse—had evidently been forming in the early period of Oppenheimer's scientific career.

In other ways, Oppenheimer and Heisenberg were rather similar, not least in their unwillingness to oppose military authority. The final irony was that after the war Oppenheimer became the victim of a group in the military who wanted him to be even more subservient to their projects, whereas Heisenberg succeeded for decades in making most people believe he had never even thought about making a nasty weapon.

Once Oppenheimer became a public figure, he could no longer afford to reveal his true state of mind as clearly as he had done in his early letters. That makes the early correspondence so valuable, for he shows himself not yet able to hide the melancholia and the turmoil of character formation that—as he put it in a letter written shortly after he had passed through the worst of it—is hardest on "men who combine a certain weakness, timidity, hesitancy of character with a quite robust vanity—or, perhaps more accurately, with an urgent desire for excellence."

The Bitter Harvest

When Oppenheimer bade farewell to his collaborators at Los Alamos shortly after their work had helped to hasten the end of the war, he was only in his early forties. The turbulent periods of youth and of service to his country behind him, he could well have been expected now to flourish into the extraordinary physicist, as his work before the Manhattan Project clearly promised. But other seeds had been also sown, and it was these that grew ferociously and strangled the rest. And it is for this surrealistic and crushing denouement that Oppenheimer will perhaps be most remembered in the future.[2]

Between the beginning of the war in Europe and the mid-1950s, three complex systems grew in the United States, each unlike any in American history: the nuclear arsenal, the science-advisory system, and an intelligence apparatus whose objective was to ferret out "security risks." Oppenheimer, a Founding Father of the first two, became a victim of the third. The core of the "Oppenheimer case" was the fight over the limits of dissent allowed a scientist working with the military—itself a part of the struggle for civilian control over national destiny, even in a democracy. For it was not only Oppenheimer who was on trial when his fitness to remain a government consultant was investigated by an Atomic Energy Commission Hearing Board under the chairmanship of Gordon Gray in 1954. The Board stated frankly, "We are acutely aware that in a very real sense this case puts the security system of the United States on trial."

Then the Board went on to define a new concept: "We believe it has been demonstrated [by these proceedings] that the Government can search its own soul and the soul of an individual where relationship to his Government is in question." While the Board members thought this was done with "full protection of rights and interests of both," what is most striking to the reader of the proceedings today is the danger inherent in the concept of the government "searching" an individual's soul. How could even the most elementary rules of due process be upheld in the pursuit of that game? If the official account of Oppenheimer's case sometimes reads as

if it were the transcript of a purge trial in the Soviet Union, it may be because the record of that trial[3] was originally not meant to be publicly accessible. The chief hearing transcript was reportedly hastily declassified
after an A.E.C. Commissioner accidentally lost his copy of the summary
on a train.

Oppenheimer was somehow a fitting hero for the tragedy that befell
him; both his strengths and his weaknesses were built on a grandiose
scale. The frail, charismatic man—a beloved teacher and an impressive
contributor to physics who set the style for students and colleagues—had
come to be both one of the most thoroughly cultivated among American
scientists and also virtually the last one who still tried, with much success,
to understand the whole range of major contemporary problems in theoretical physics. To the end of his days he could rise in almost any seminar on current work to point out a significant connection or improvement. Usually he did so with lucid elegance, but sometimes with
withering disdain. Both of these traits became of consuming interest to
those who had to judge him against the requirements of the security
system. For while he very often was right, he was sometimes wrong
and destructive. And with all his erudition he never overcame an early
political naiveté. To the men who took it to be their task to search his
soul, their quarry must have seemed a strange mixture: part Ulysses,
part Faust, part Prince Myshkin.

When Oppenheimer left his sheltered joint position at the University of
California-Berkeley and the California Institute of Technology physics
departments to accept the leadership of the Los Alamos Laboratory in
1942 at the urging of General Groves, he had already helped to make the
United States a center for advances in contemporary physics. Now he put
every ounce of his being into the race against the presumed German effort.
General Groves, who stood by him to the end, had chosen Oppenheimer
well. "Oppy" could get almost anyone to join him. Later, on the various
policy and advisory boards in which he was immersed immediately after
the war, the same ability to make a group work at its best and to provide
persuasive arguments and summaries was attested to by such colleagues
as James B. Conant, I. I. Rabi, Cyril Smith and Vannevar Bush. As one of
them said later, "Robert would put things so clearly that after he had spoken, no one else felt there was any need to do any thinking."

It was just this that came to worry the Air Force and particularly the
Strategic Air Command (S.A.C.), when Oppenheimer's policy advice
seemed to go against them. At the A.E.C.'s Gray Board Hearing, Air
Force General Roscoe Charles ("Bim") Wilson ("I am first of all a big-
bomb man"), who was sent there explicitly on orders from the Chief of

Staff of the Air Force, explained to the judges some of the reasons for his mounting worries about Oppenheimer:

1. Oppenheimer's support for the Acheson-Lilienthal-Baruch postwar plan to internationalize atomic energy controls—"when the U.S. had a monopoly."

2. Oppenheimer's doubts about the early feasibility of nuclear-powered aircraft (correct doubts, as it turned out), in contrast to his apparently smaller doubts concerning the possibility of nuclear-powered ships—just when the Air Force was fighting the Navy for budget and dominion.

3. Oppenheimer's approach to the hydrogen-bomb project, an approach that showed "more conservatism than the Air Force would have liked."

4. Then, revealing his most pertinent worry, the General confessed: "I would like to say that the fact that...he is such a brilliant man, the fact that he has such a command of the English language, such national prestige, such powers of persuasion, only made me nervous, because I felt if this was so it would not be in the interest of the United States, in my judgment. It was for that reason that I went to the Director of Intelligence to say that I felt unhappy."*

In 1954, the "fantastic opinion" on trial was the suspected reason behind Oppenheimer's view of 1949—in which he had been joined by virtually all his colleagues on the General Advisory Committee (G.A.C.) that reported to the A.E.C. and to the President—that no crash effort should then be initiated by the United States to produce the hydrogen bomb, a new weapon much desired by S.A.C. Oppenheimer's panel had been asked, soon after the September 1949 announcement of the first Russian A-bomb explosion, to provide the A.E.C. with "advice and assistance on as broad a basis as possible." At its meeting on 29 October 1949, Oppenheimer, Enrico Fermi, Conant, Lee DuBridge, Rabi, Hartley Rowe, Cyril Smith, and Oliver Buckley (with Glenn Seaborg absent) reviewed the arguments for and against the United States initiating the awesome next step in the arms race. There were fundamental doubts whether the "miserable thing will work," as Oppenheimer expressed it. Also, the weapons race had already reached overkill level, with construction of a half-megaton fission bomb authorized. And the stock of plutonium would not be sufficient both for new developments and the current weapons-diversification plans.

Above all, it seemed to these men dangerous to go on to produce a superweapon having, unlike the A-bomb, no natural upper limit to its de-

---

* In a brilliant essay, Giorgio de Santillana noted a similar charge among the accusations against Galileo: "He reminded me that Galileo wrote exquisitely, and had a marvelous capacity for persuading people of whatever he wanted to, and there was a danger that through his influence some fantastic opinion might take hold among those Florentine wits, which are too subtle and curious." Giorgio de Santillana, *Reflections on Men and Ideas* (Cambridge, MA: M.I.T. Press, 1968), p. 131.

structive power, and to do so in a unilateral escalation of the arms race—without having first allowed an honest effort to be made at mutual arms control. Fermi and Rabi condemned an initiation of the development of such a weapon as "wrong on fundamental ethical principles"; the others on the G.A.C. saw this moment as the unique opportunity "to provide by example some limitations on the totality of war, and thus eliminating the fear and arousing the hope of mankind."

Admittedly, at the time the prospects for arms-control agreements with Russia were not good. Oppenheimer himself, in his work on the Baruch plan, had seen that much earlier. In fact, even within the U.S. there was no forum for scientists who wanted to discuss the technical aspects of arms control proposals, and when three individuals (Rabi, Oppenheimer, and Charles Lauritsen) felt it wise to meet informally for a working seminar, the Air Force formed the absurd idea that their gatherings were part of a hostile conspiracy. In the same vein, Oppenheimer's principal accuser, William Borden, saw only one explanation for his opposition to the development of the H-bomb: Oppenheimer, Borden said, was very probably "an agent of the Soviet Union."

The "fantastic opinion" that one does not immediately have to rush ahead to make any weapon that can conceivably be made constitutes a Principle of Abstinence, whose need was understood earlier and more naturally by some scientists and engineers than by the Joint Chiefs, the A.E.C., and the security system. When the majority of the Hearing Board, after a month of excruciating sessions in 1954, decided not to reinstate Oppenheimer's security clearance—an action later reaffirmed and expanded in a series of subsequent steps within the A.E.C., each with its own breathtaking twists—the reason the Board gave turned out after all not to be that there was evidence to doubt Oppenheimer's loyalty and discretion. On the contrary, the Board volunteered that "Oppenheimer is a loyal citizen," and that he "seems to have a high degree of discretion, reflecting an unusual ability to keep to himself vital secrets." Despite all the material in the investigation reports of the F.B.I. and in reports standing several feet long in the A.E.C. files, despite a combing of fifteen years of Oppenheimer's life—during most of which he was under constant surveillance, his movements watched, his mail and telephone calls checked, his professional monitors aided by help volunteered by personal enemies—the worst indiscretions that could be found were Oppenheimer's foolish lies a dozen years earlier regarding the Chevalier incident—an event well known to the A.E.C. when Oppenheimer had been cleared in previous investigations.

But none of that was in fact what this trial was all about. Otherwise, it might have been far simpler for the A.E.C. to let its contract with Oppenheimer lapse without renewal, since it was to expire anyway within weeks. Rather, what had to be done in view of Oppenheimer's dangerous notions about the H-bomb project was, consciously and symbolically, to remove a very visible but unsuitable scientist from the corridors of power. This was clearly stated by A.E.C. General Manager Kenneth D. Nichols (previously second in command under General Groves, and in that position one of the several recipients of Oppenheimer's tongue lashings that came back to haunt him later). Nichols said that if the security clearance were not revoked, other agencies of the government might use Oppenheimer, and the glow of his influence in the government would remain. It also did not help matters that in a Congressional hearing in 1949 Oppenheimer had deeply offended Lewis Strauss by ridiculing his opposition to the shipment of radioisotopes to friendly countries; later Strauss was in the position, as A.E.C. chairman, to oversee the Oppenheimer security hearings.[4]

The only serious charge on which the Gray Board convicted Oppenheimer was that his "conduct in the H-bomb program was sufficiently disturbing as to raise a doubt" whether his continued participation was consistent with "the best interest of security." An example given of the suspicious "conduct" was that after the presidential go-ahead for the H-bomb program was made public on January 31, 1950, Oppenheimer "did not make known that he had abandoned" his anti-H-bomb views, and therefore "undoubtedly had an adverse effect on recruitment of scientists" for the program. "Enthusiastic support on his part would perhaps have encouraged other leading scientists to work on the program."

Ordinary loyalty and discretion had not been enough. As the Board put it, "absolute loyalty" would be required for clearance. A scientist's right to express dissent from current national policies, even during secret advisory meetings, or his right at least not to wax enthusiastic about policy decisions which seemed to him unwise—these rights were, to the "judges" of the A.E.C., fantastic notions, particularly if such lack of enthusiasm appeared to be based on moral grounds.

Here was irony indeed, for it was in large part Oppenheimer's success at Los Alamos that had converted the image of the scientist from an abstract character living in an ivory tower into a major force in policy matters. And it was also he who was among the first to realize that a "technically sweet" job can nevertheless be diabolical. To the uncomprehending General Counsel who played the role of prosecuting attorney in the Gray Board hearings, Oppenheimer said gravely, "I felt, perhaps quite strongly, that having played an active part in promoting a revolution in warfare, I needed to be as responsible as I could with regard to what came of this revolution." But the Board held that Oppenheimer had gone "beyond the proper function as a scientist." What emerged from this decision was the doctrine that the function of a really loyal scientist is enthusiastically to participate in an uncontrolled arms race.

Not the least of the ironies was that in 1949 President Truman had almost immediately overruled the advice of Oppenheimer and the rest of the G.A.C. and ordered the H-bomb project to go ahead. Then Edward Teller, who from the beginning had been caught up in his irresistible urge to develop the ultimate weapon, showed how the bomb could be made more simply than anyone had thought, and now Oppenheimer went along with everyone else.

Moreover, Oppenheimer's heretical suggestions to promote arms control and to reduce reliance on superweapons did eventually become part of avowed national policy. Nor did the Gray Board decision really do its job; for over the years the doctrine that there is an enforceable breach between narrow technical advance and the rest of the spectrum of human concerns has become widely discredited among scientists, partly owing to the fact that it simply did not work, either in military or in nonmilitary technology. On the contrary, more and more scientists, with considerable courage in the face of great pressure, put their reputation on the line by objecting publicly against military objectives, from antiballistic missiles to chemical and biological warfare strategies and "Star Wars," as well as against nuclear weapons tests and "defoliants."

In the last of its "general considerations," the Gray Board majority dealt with the "role of scientists as advisers in the formulation of Government policies." They worried that scientific experts may have "vastly greater" influence than the individual citizen, and hence they set down guidelines for their proper behavior. The most remarkable suggestion was that a good scientist-adviser must have "genuine conviction that this country cannot in the interest of security have less than the strongest possible offensive capability in a time of national danger." Such a rule could of course leave military policy entirely in the hands of hawks working on doomsday machines. It was perhaps this ghastly vision that Oppenheimer glimpsed on the threshold of the H-bomb Age, and that planted the seed for this fantastic notion of forming an independent judgment of how technology should be used.

It has often been said that Oppenheimer himself was his worst enemy in this complex case. It is also still most dispiriting to see how ineffective Oppenheimer was under fire, how he acquiesced in his own destruction. As he said later, "I had very little sense of self" in the hearings, and to at least one observer of the proceedings he gave the impression of playing out the Greek tragedy of a man's downfall.

Perhaps the psychological toll of years of living in an unending weapons race had been too much. Perhaps it was not merely theatricality that made him say "I have blood on my hands" when Acheson once brought Oppenheimer into Truman's office. (As Acheson later told again with relish, Truman responded to Acheson with manly invective, "Don't ever bring that damned fool in here again. He didn't set that bomb off—I did. This kind of snivelling makes me sick.") Perhaps he had just clung too long to the hope for a rational order while embroiled in an insane procedure. David Lilienthal, when browbeaten in the usual manner while testifying on behalf of Oppenheimer during the hearing, saw the chaotic procedure for what it was. He discovered that Oppenheimer's lawyers, not having clearance to see crucial documents in the case, had to leave the room, and he commented, "when I saw what they were doing to Oppenheimer, I was ready to throw chairs." Four years later the new General Counsel of the A.E.C. was reported to say, on reviewing the case, that it had been "a punitive, personal abuse of the judicial system."

But Oppenheimer, the architect of monstrous weapons, was not a man to throw chairs. In any case, it would not have changed anything. Yet it is a measure of the tragedy that today, rereading the transcript of the hearing, one almost wishes he *had* thrown some chairs about. The story calls for a climax, a dramatic act to mark the end of the road that had been laid out for Oppenheimer by the prevailing conceptions of the scientist-adviser and of national security, and by his own acquiescence in the hearing proceedings.

But the most obvious unresolved puzzle is, what deprived Oppenheimer of his self-respect in this trial? As Philip Stern put it, "What prompted this man, so often accused of being self-willed, arrogant and disdainful of the views of others, to be so self-contemptuous in his own defense?" Even one of the A.E.C. Commissioners, Thomas Murray, who later condemned him for his "associations" and voted to uphold the Gray Board conclusions not to grant Oppenheimer's clearance, saw nothing wrong with the infusion of "moral reasons" in debates concerning technical decisions; Murray said of the scientist: "Like every man, he ought to be alert to the moral issues that arise in the course of his work."

The fact that the "security system" at the time took it upon itself to condemn scientists precisely for this attitude could in the long run serve to endanger the security of the nation. But the full explanation why it could so easily visit such a trial on this particular scientist, and bring him down with the help of his own self-contempt or identity confusion—that surely has some roots in the earlier, personal trials of the turbulent years of his youth. Apparently those wounds, which had never healed completely, gaped open during his demeaning collision with an authority system that denied the very basis of his self-respect—as a scientist, a patriot, an intellectual, and an adviser at the highest political level who also tried to face the moral issues.

References

1. Alice K. Smith and Charles Weiner (eds.), *Robert Oppenheimer, Letters and Recollections* (Cambridge: Harvard University Press, 1980). All quotations in the next few pages come from this book.
2. Again, among the many story books concerned with that part of the story, there was one that seems to me most alert to the contrasting forces involved, and from which I shall occasionally quote in the pages that follow; it is Philip M. Stern, *Security on Trial,* with the collaboration of Harold P. Green, and a special commentary by Lloyd K. Garrison (New York: Harper & Row, 1969).
3. U.S. Atomic Energy Commission, *In the Matter of J. Robert Oppenheimer* (Washington, DC: U.S. Government Printing Office, 1954).
4. See Philip M. Stern, *The Oppenheimer Case* (New York: Harper & Row, 1969), pp. 128–130.

Percy W. Bridgman, Physicist and Philosopher

Percy W. Bridgman (1882–1961) won the 1946 Nobel Prize in Physics for his investigations of the changes in materials subjected to extremely high pressures. He is also widely remembered for his contributions to the philosophy of science, which changed the intellectual landscape of many scientists. His philosophy of "operationalism," first elaborated in *The Logic of Modern Physics* (1927), attempted to brush away undefinable elements by proposing that concepts should be defined in terms of experimental procedures, or operations—an approach that influenced economics and psychology also.

I had the privilege of knowing Bridgman, first as his dissertation student in experimental high-pressure physics, and then as a colleague. It seems to me that his life and work were constructed along about half a dozen ordering ideas.

1. An inductive style of research that brought brilliant experimentation to an astounding level

In this regard one is reminded of Faraday: in the simplicity of the equipment, and the thoroughness of detail of the research accomplished with it. When I first entered Bridgman's laboratory, fresh from doing war-time research with advance pulse circuits, I was astounded to see nothing but DC electrical measurements of resistances and potentials, and Kelvin Bridges instead of electronics. Yet the experiments—on tiny samples the size of a nail paring, conducted inside small presses that were themselves inside bigger presses, and all the information on the changing physical properties coming out on one wire to an old galvanometer—were top-drawer. The work Bridgman did was always personal, hands-on, faster and with more data points per day than seemed humanly possible. He said "physics is the quantitative exploration and analysis of physical experience." Those words meant what they said: quantitative, exploration, experience. He was guided not by fashionable new theories but by thermodynamics, and the prospect of huge areas of new physical phenomena to be conquered. He gave the impression of wanting to harvest fields upon fields of new data while keeping his eye on yet more remarkable ones constantly appearing on the horizon as he increased the pressure range.

2. No barriers

Like other scientists I have studied, Bridgman saw no fundamental difference between science and philosophy if properly conceived, nor really between his life and his work, nor between public and private science. To him there was no distinction between a novice with interesting ideas and a willingness to work and to take risks and an elder statesman of science on a courtly visit. He seemed to be curious about every subject in science and outside, but by the same token, he was never satisfied that he had really exhausted any deep question. He would not have been surprised to learn that his own extensive work (seven technical books and some 200 scientific papers, plus seven more books and about 60 papers on philosophy and the social study of science) is useful today in a great variety of natural sciences—physics, chemistry, geology, engineering, even in biology, as in the study of the properties of protein under high pressure—as well as in psychology, economics, philosophy, and history of science. As a result, the Bridgman archive at the Harvard library, which contains more than thirty linear feet of lab notes and correspondence, is, according to the archivists, now among the most constantly used of all the personal collections in Harvard's holdings for scholarly research.

3. Loyal and stubborn persistence

Bridgman stayed with his university from the day he entered as a freshman in 1900 until he retired from his professorship 54 years later. He persisted in the same field of research throughout, and saw it grow in power as he patiently drove the range of attainable pressures in his laboratory from a few thousand atmospheres at the start, to about 20,000 in 1910, to 50,000 in the mid '30s, and ultimately to over 400,000. The same persistence came out in the short run as in the long. A famous example is a sentence from one of his scientific papers, describing how to make a thin hole as narrow as the lead in a pencil, but in a big block of very hard steel: "It is easy, if all precautions are observed, to drill a hole...17 inches long, in from 7 to 8 hours."

Anyone who knew him can immediately see him in his well-used lab coat, at the lathe or at the fore-pump of the press, pumping up the pressure, always by hand, undeflectable by anything while taking data. I can attest to that. In the fall of 1946, I was in the workshop, constructing apparatus for a thesis problem to which he had agreed as long as I would not bother him too often. The telephone rang, and I happened to be the one to take it. A voice from the Associated Press asked me to bring Mr. Bridgman to the phone. Bridgman, of course, did not tolerate a telephone in his laboratory. I explained it was impossible to get him out while he was taking data. The voice at the other end, now quite agitated, demanded an interview with him because the news had just come over the wire that Bridgman had won the Nobel Prize. Of course I ran into Bridgman's lab with the great news. Not missing a single stroke, he continued at the pump and said quite simply, "Tell them—I'll believe it—when I see it." Here indeed was the operational approach in action.

4. A lifelong struggle for personal intellectual clarity

Plato said that "clear ideas drive away fantastic ideas;" but in Bridgman the need to clarify his ideas in honest self-evaluation was not merely therapeutic, it was a biological necessity. He had an ethos of lucidity. In 1938, at the peak of his powers, he wrote, "As I grow older, a note of intellectual dissatisfaction becomes an increasingly insistent overtone in my life. I am becoming more and more conscious that my life will not stand intellectual scrutiny, and at the same time my desire to lead an intelligently well ordered life grows to an almost physical intensity."

So, for Bridgman, another all-too-frequent barrier is removed—that between moral and intellectual issues. Compelled by this insistence to be clear about one's stand, he published a manifesto in *Science* in which he announced closing his laboratory to visiting scientists of totalitarian countries because he saw they had subordinated their loyalty to science to the demands authoritarian governments might make on their work. But we also see in the quote above a mainspring, a force driving several apparently disparate achievements. His attractions to high-pressure physics seem to have started with his attempt to give instrumental meaning to the concept of high hydrostatic pressures inside a sealed enclosure. He never tolerated the use of black boxes that required the experimenter to believe a manufacturer's calibration or circuit diagram. His work on dimensional analysis was in the nature of a self-interrogation. His book on thermodynamics starts with the question, what are the most basic variables to choose and how do we define them.

Bridgman may not have read Charles Peirce's essay of 1878, "How to Make Our Ideas Clear," but he had a deeply rooted affinity with the American empiricist philosophers. Here, too, was the source of his sympathy with the philosophy of the displaced Vienna Circle, as brought by refugee scientists and philosophers who began to come to the United States in the 1930s.

The Logic of Modern Physics (1927) was his first major contribution to the critical analysis of the foundations of physics. To this day, anyone with scientific interest who reads it for the first time finds it an electrifying experience. As is true for many of us, I clearly remember the exact place in the library where, as a student, I happened upon the book and then found myself standing there, reading the first chapters right away. Key phrases burn in one's memory: "In general, we mean by any concept nothing more than a set of operations; the concept is synonymous with the corresponding set of operations." Or: "The true meaning of a term is to be found by observing what a man does with it, not by what he says about it." Or: "If a specific question has meaning, it must be possible to find operations by which answers may be given to it. It will be found in many cases that the operations cannot exist, and the question therefore has no meaning."

All the traits of Bridgman's mind were there: the lucidity of style, the lack of philosophical self-consciousness, the uncompromising drive to the foundations. It was a message of immense power for scientists, at least in the English-speaking world; and as in many such cases, the explanation of its power is not that the work brings to the reader a message never thought of before, but rather that it lays open, with clarity, what the reader has been trying to formulate on his or her own. The most fruitful way of understanding Bridgman's philosophical writings is to see them as a record of continual self-discovery. That is why they have also been such an effective tool of self-recognition for other scientists.

It is not surprising that Bridgman had to clarify for himself philosophical problems before he could feel satisfied with his solution of scientific problems. For the typical task of science, to put it very simply, is to relate the world of appearances, or data, and the world of conceptions, or ideas. And the usual way scientists do this is to construct an arch that starts from the ground of direct observation, then rises up to the region of theory, or concepts and other thought constructs, and then comes again back to the level of the immediately given, to a new set of verifiable observations.

The question that always nags the honest mind is this: how can one be sure whether or not the arch connecting initial and final data is not mere fantasy? Roughly, Bridgman's answer was that there are two tests. There must be formal connections among the thought elements in the upper region; they must be able to stand the tests of logic and of "mental" or "paper-and-pencil" operations. And second, each of the physical concepts used must, in principle, be supported from the world of experience below, by having meaning in terms of physical or instrumental operations. This is how the operational attitude determines the shape and direction of a possible theory and type of data relevant to it at each end. But this process of construction is, of course, not restricted only to science, and Bridgman's operationalism spread far beyond the physical sciences. In his straightforward way, he wrote: "It would doubtless conduce greatly to clarity of thought if the operational mode of thinking were adopted in all fields of inquiry." Even so, the critique and elaboration of his conceptions by other scholars sometimes astonished Bridgman. Once, at the end of a long symposium with philosophers, he said:

> As I listened to the papers I felt that I have only a historical connection with this thing called "operationalism"....I have created a Frankenstein, which has certainly got away from me. I abhor the word *operationalism*...which seems to imply a dogma, or at least a thesis of the some kind. This thing I have envisaged is too simple to be dignified by so pretentious a name; rather, it is an attitude or point of view generated by continued practice of operational analysis....
>
> The date usually associated with this is 1927, the year of publication of my book, *The Logic of Modern Physics*, but preparation for this in my own thinking went back at least to 1914, when the task of giving two advanced courses in electrodynamics was suddenly thrust upon me....The underlying conceptional situation in this whole area seemed very obscure to me and caused me much intellectual distress, which I tried to alleviate as best I could....The dimensional situation proved comparatively simple, and I was able to think the situation through to my own satisfaction—an experience that perceptibly increased my intellectual morale.[1]

In this quotation there is, I believe, a clue to Bridgman's key motivation. It is the sentence, "I was able to think the situation through to my own satisfaction." To clear his own ever-active mind of intellectual disquietude—that was a basic driving force, whether designing an experiment or analyzing the foundations of thermodynamics or of society. He never bothered to answer seriously the charge of philosophical solipsism, but I think that the answer has to be this: One cannot, by any *other* criterion, hope to get maxims that are more generally valid than those one can get by letting questions rise up in a mind that habitually subjects itself to exceptional intellectual disquietude and that habitually is given to uncompromising self-examination.

5. Uncompromising individualism

With an idiosyncratic ferocity, Bridgman was responsible to himself first and last. His most critical audience was himself. Thus only a small handful of his papers are collaborations, and he accepted fewer than twenty doctoral students in his whole career. When Bridgman took down and worked out data, he usually had one more significant figure in his notes than he would publish. I don't think this was due only to caution—he also enjoyed keeping something for his own pleasure, and future harvesting.

His individualistic approach allowed him to do with remarkably little research money, apart from the part-time support of two helpers, the machinist Charles Chase and the data-taking assistant Leonard Abbott. As the pressure range and complexity of the presses increased, he feared that the new style of bigger science would encroach on the scientist's personal liberty. In one of his last articles, he wrote, "It is a serious disadvantage that experiment in this field is [now] so expensive that it almost demands governmental support, with all the unwelcome complications of such support."

First and last, his science was his individual struggle to discern "the way things are," which is, in fact, the title of one of his last books. Indeed, he told me he had preferred to call this book, "The Way It Is," but, he said, "the publisher doesn't like it; however, I am still not sure that 'things' really exist." To him, the mark of a scientist was the willingness to devote oneself fully to finding out the way it is, regardless of the consequences.

Thinking about him in this way, I am certain that the requirement of intellectual integrity demands that one must also speak in plain language of his tragic death on August 20, 1961. Already in an essay published three decades earlier, entitled "The Struggle for Intellectual Integrity,"[2] Bridgman dealt with the question of choosing death if one's future is likely to be predominantly painful. We know that this choice came to him with great suddenness in the eightieth year of his life, and we know also that he was not a man to think one thing and do another. In July 1961, after careful diagnosis by undoubted authorities, he learned that, as he put it in one of his last letters, the "disease has run its normal course, and has now turned into a well-developed cancer for which apparently nothing can be done....In the meantime there is considerable pain, and the doctors here do not offer much prospect that it can be made better." Of course it was not merely a matter of pain, which he bore with such grit that he could still finish the manuscript of a new book in his final year. More important, I think, Bridgman could not conceive of himself as unable to *act*. And so he added this sentence: "I would like to take advantage [*sic!*] of the situation in which I find myself to publish a general principle, namely, that when the ultimate end is as inevitable as it now appears to be, the individual has a right to ask his doctor to end it for him."

That was, however, not a possible course; and so he wrote a laconic, two-sentence note on the final day: "It isn't decent for Society to make a man do this thing himself. Probably this is the last day I will be able to do it myself. P.W.B."

A day after his death, Harvard University Press received the last thing he was working on: It was an index for the forthcoming seven-volume collection of his complete scientific papers.

Near his beloved garden in Randolph, New Hampshire, which he had planted again that spring, his ashes are buried near a tree that stands by itself, looking out to Mt. Adams.

References

1. Percy Bridgman, "The Present State of Operationalism," in Philipp G. Frank, ed., *The Validation of Scientific Theories* (New York: Collier Books, 1961), pp. 75–76. The book was first published in Boston by Beacon Press in 1954.
2. "The Struggle for Intellectual Integrity" was published in *Harper's Magazine* in December 1933. It was reprinted in *Reflections of a Physicist* (New York: Philosophical Library, 1955), pp. 361–379.

I. I. Rabi, Citizen-Scientist

T he summer of 1940 was a desperate and decisive time for the free world. The Nazi armies controlled most of the European continent. Their bombers were pounding England, and their U-boats were devastating the seas. Desperately needed was a short-wave, high-intensity radar system to "see" enemy aircraft, ships, and ground installations.

In the British laboratories, the embryonic heart of such a system, the cavity magnetron, had recently been perfected. On October 6, 1940, the device was brought to the United States, where it galvanized activity among American scientists and engineers. Less that two weeks later, a microwave laboratory was initiated at MIT, the "Rad Lab," to transform the magnetron somehow into prototypes of working radar systems.

The scientific and engineering obstacles were immense. But within a few months the first prototypes were being battle-tested, and when the United States was pulled into the war in December 1941, it had not only the basis for a multitude of specialized radar systems but also a model for mobilizing brains during a war. Radar on aircraft, shipboard, and land was crucial in breaking the back of the Axis offensive. As Lee DuBridge, director of the Rad Lab, put it: "The bomb ended the war, but radar won the war."

The detailed story of radar in World War II, unlike that of the nuclear weapons, is being appreciated only slowly by historians. (Even Henry Guerlac's official history of the Rad Lab had been lying on the shelf unreleased for four decades until the American Institute of Physics published it in 1987.[1]) But it is clear that a central figure in the development of radar was I. I. Rabi, later the dean of American physicists and until his death in 1988 one of the most savvy and engaging scientists of our time. His role as chief researcher on the magnetron, as talent scout *extraordinaire*, and as liaison with the military was just one of the chapters in his life, told succinctly, with excitement as well as good scholarship, in John S. Rigden's book, *Rabi: Scientist and Citizen*.[2]

Isidor I. Rabi's is an American tale in the tradition of unlikelihood mixed with seeming inevitability. Brought as an infant from Galicia, Rabi grew up on the lower East Side of New York City, and, later, in Brooklyn, on one side hemmed in by poverty, but expanding on the other into the culture and religion of Orthodox Judaism. His special gift of clear and captivating speech was evident early. Though always small of stature, he made it on the tough slum streets by fascinating the bully-boys with Bible stories.

The local public library was his refuge, a place for his self-education but eventually the cause of a familiar crisis. A little book on astronomy opened his eyes to the Copernican solar system with its clockwork regularity of planetary motion. The simplicity and beauty were overwhelming, scuttling his biblically-based belief that the sun's risings and settings are caused by God's constant operations. Young Rabi announced to his pious, long-suffering parents: "It's all very simple: who needs God?" He was evidently one of those fiercely independent kids: when time came for his bar mitzvah, it was held, at his insistence, in his home instead of the synagogue, and he delivered a speech on "How the Electric Light Works." Yet all his life he remained God-struck, rather like Einstein, who called himself an irreligious believer.

Winning a scholarship in 1916, Rabi studied chemistry at Cornell University, but he remained fascinated by history and particularly by the history of science. On graduating, he discovered a fact so easily forgotten today: At that time, neither he nor his Jewish classmates could obtain job offers from the chemical industry, nor were Jews welcome in faculty positions in colleges and universities. The best he could find was a small job in a company lab, then as a bookkeeper. For three years, he and his closest friends were floating, going almost every night to the New York Public Library to study science on their own. Finally, excited by the new discoveries in atomic structure, he returned to Cornell, this time to study physics. He had at last reached the field in which his intuition and insights would operate superbly. Moving in 1923 to Columbia University for graduate work—chiefly to be near Helen Newmark, a young lady he had met recently and whom he would marry as soon as he had his Ph.D.—he also found the university that was to be his home base for the rest of his career.

Those were heady times in physics. Otto Stern and Walther Gerlach had shown that atoms traveling in a molecular beam line up as they pass through a magnetic field, taking only one of two positions in space. This evidence of "space quantization" was completely contrary to the early form of quantum physics, not to speak of classical ideas. Out of such puzzles developed the successive versions of quantum mechanics in the second half of the 1920s. But one had to be in Europe to be in on the excitement. With a fortunate fellowship, Rabi, like many other Americans at the time, could now go for a year to acquire the taste of the new physics. He explained, "We knew the libretto, but we had to learn the music."

For Rabi, perhaps the most decisive turning point began as yet another accidental encounter. Visiting Otto Stern's molecular-beam laboratory in Hamburg, Rabi was almost casually invited to stay on beyond his fellowship and try his hand at a new version of the molecular-beam experiment that had occurred to him. Rabi knew that back home the outlook for getting a job as a physicist was just as bleak as it had been for him as a chemist. He was in no position to refuse the honor.

Previously, Rabi had seen himself as a theoretical physicist. Now he had to teach himself the necessary experimental skills. Within a year, he was able to show the merit of his method of demonstrating space quantization. With it, he also established in his own mind his credentials as scientist at a time when European hosts such as Arnold Sommerfeld tended to look down on visiting Americans. Rabi realized that "what we needed were the leaders," and he made a compact with some of his American friends there to put an end to the second-class status of their science when they returned home.

That opportunity came in 1929, with an invitation to join the Columbia University faculty. Ironically, it was on the recommendation of Werner Heisenberg, who had met Rabi in Germany and had been impressed by him. The rest of the story—well told by Rigden, himself a physicist, who had full access to the documents and to Rabi himself—shows how intertwined have been the rise of science in American and the tumult of world history in the last six decades. Step by step, we see here how Rabi's instinct for the basic theoretical questions and for brilliant shortcuts in experiments led him to build up a pioneering research team at Columbia. His development of the magnetic resonance method, allowing precise measurements of the properties of atomic nuclei and fundamental physical constants, brought him the Nobel Prize in 1944.

As it turned out, this work served also as a providential preparation for his war effort on radar systems. In addition, he would slip out of the Rad Lab to serve as adviser to Robert Oppenheimer and his colleagues at Los Alamos. Again and again, Rabi's common sense on scientific and science-policy matters was crucial, or when not followed, proved the contrary decision to be flawed. For example, it was he who persuaded Oppenheimer not to let the Los Alamos labs be militarized. (Oppie had already ordered his Army uniform as lieutenant colonel.)

After World War II, Rabi worked assiduously to keep scientific research, as well as organizations such as the Atomic Energy Commission, under civilian control, to defuse the threatening nuclear-arms race, and to reestablish sound contacts with scientists abroad. Together with Enrico Fermi, he opposed the all-out effort to make a thermonuclear superbomb, regarding the stockpile of atomic weapons sufficient for deterrence purposes. In his official report, he wrote," We think it wrong on fundamental ethical principles to initiate a program of development of such a weapon." Instead, "it would be appropriate to invite the nations of the world to join us in a solemn pledge not to proceed in the development or construction of weapons of this category." Privately, he added: "I don't mind dying; my ancestors did that. What I do mind is the destruction of civilization.... This is the holy thing which they are violating by pushing in the direction of an annihilating war."

During the war years, many a scientist had been propelled practically overnight from some parsimonious and sheltered academic position into the highest councils and onto the world's stage. To keep one's sense of identity secure despite this transition was no easy matter. As noted earlier in the book, Oppenheimer practically destroyed himself during the hostile hearings on his A.E.C. security clearance in 1954, confessing later that he had lost his "sense of self" in those proceedings. Not so Rabi, who was called to testify on Oppenheimer. He went on the offensive against the badgering interrogators, dramatically putting the hearing into perspective by insisting that in such matters it is the achievement of a whole life that should be weighed: "What more do you want? Mermaids?"

Rabi's own sense of self was evidently made of cast iron, and he had never been afraid of saying plainly what was on his mind. Looking back near the end of his distinguished career, he was not cheered by what he saw. On November 1, 1986, speaking to a meeting of science educators at his beloved Columbia University, he again put his beliefs on the line:

> Science, for its progress, its understanding and so on, depends so much on truth and on clarity. And the question is: Are we living in a society where truth and clarity are an important element of our culture? And I must say, that this administration—and I speak as a lifelong Republican!—is just morally bad for young people. You have the highest administrative officials telling lies that are not punished. That is a terrible moral situation: where you have, on the very top, show business! And the truth is not there. You have a situation, in the cherished program of the president, namely the 'stars wars' program, where a large fraction of the top scientists of the United States are opposed to it, on moral, but much more on technical

grounds. And they speak out; but there is no answer; they are not answered at all; they are just brushed aside.

So I am making this proposition to you: that the teaching of science is so much more than the imparting of knowledge, or imparting a certain method, but goes back to the really basic principles on which the Republic was founded: It is to try to give the student a vision of the beauty of truth.

References

1. Henry E. Guerlac, *Radar in World War II* (Los Angeles: Tomash Publishers; [NY]: American Institute of Physics, 1987).
2. John S. Rigden, *Rabi: Scientist and Citizen* (New York: Basic Books, 1987).

Feynman's Adventures

I. I. Rabi once told me that he found that young people are led into science by struggling either with their radio set or with their God. For Richard Feynman, Nobel Prize physicist and one of the brightest stars in that select company, it was evidently not the deity. His madcap first volume of autobiographical writings[1] is a collection of irreverent, amusing and sometimes slightly raunchy reminiscences of his life's journey from Far Rockaway, New York to Pasadena, by way of MIT, Princeton, Los Alamos, Cornell, Stockholm, and Rio de Janeiro. Feynman tells us in unsparing and sometimes hilarious detail what this "curious character" has discovered about himself at each stage.

First, there is indeed the little boy—his curiosity nurtured by the clever teaching of a sensitive and beloved father. Feynman learns that he can repair a radio set by *thinking* and that he thinks best by visualizing what makes it go, step by step. He easily memorizes shortcuts in mathematics, he loves to guess and experiment, to play mischievous pranks and improvise. Most of all, he enjoys a battle of wits: but—the stereotypical boy-scientist—he has no idea at all, just then, what to do either about girls or about the college requirement to get some "culture." When a worthy matron at Princeton asks him whether he would like cream or lemon in his tea, the Huckleberry Finn in him replies: "I'll have both." Her astonished response becomes the book's main title.

But Feynman is not joking. To be sure, he is having fun all the way, even though one begins to suspect by and by that there is a dead-serious message hidden under the self-derogatory tales. He falls in love with biology but gets out before he is trapped in it: like all really good people who can do a dozen things very well, he smells out early where his best talent lies and what he must reluctantly leave alone. And on the side, he lets his restless curiosity lead him on wherever it will: He figures out how to open the super-secret safes at Los Alamos. He devises experiments to see whether ants can be taught geometry. He learns how to play percussion instruments and, in disguise, struts down the main drag of Rio during Carnival, performing with a Samba school from a favela. In a Chaplinesque scene, he tries hard to learn how to swagger up to a tough bar, how to hold his liquor, and how to win a fistfight. He experiments with self-induced hallucinations and "out-of-body experiences" while suspended in a sense-deprivation tank.

Almost by accident, he becomes an expert lecturer on Mayan astronomy. He signs up for a correspondence course to learn how to draw, and does well (perhaps another aspect of his power of visualization). Discovering his pleasure and skill in drawing from life models, he hits on a simple way of saving the cost of hiring them—a topless bar near home, where the professor finds himself several times a week, sipping his 7-Up, drawing away between doing some physics or preparing a lecture. (He tells us that his wife, who is English, "has a good attitude" about this: she simply regards the bar as his equivalent of a British club.)

But all this—and much more levity like it, such as fooling the censors of Los Alamos, testing whether humans can act as bloodhounds, finding the key to the hearts of showgirls and to the pockets of gamblers in Las Vegas, playing bongo drums to accompany ballet performances—all this is simply diverting, in both senses of the word. Feynman, who is as allergic to any solemn philosophizing as was Voltaire's Candide, hides the mainspring of his motivation behind those anecdotes. Although he doesn't tell it here, he used the same device in his Nobel Prize lecture in 1965, the first time he gave a lecture wearing a suit. There he told the undoubtedly startled king and his royal family, "I shall include details of anecdotes which are of no value scientifically, nor for understanding the development of ideas. They are included only to make the lecture more entertaining." But having said this much, Feynman went on to give one of his legendary lectures, cast in his light-hearted storyteller's style but utterly devoted to the new ideas in physics and to ways of getting, proving, and using them.

The turning point in young Feynman's career came when John Wheeler accepted him as a graduate student at Princeton. It was a fortunate meeting of the unkempt and the urbane. Wheeler possesses unlimited patience as well as unlimited imagination and daring. Feynman quickly blossomed, absorbed Wheeler's famous dictum, "Make the mistakes as fast as possible," and began to sift out the good ideas, often superbly graphic, which have become part of every physicist's tool kit. For example, Feynman showed that the probability amplitude, which can be thought of as traveling wavelike with an electron as it goes from one point A in space-time to another point B, is the result of the superposition of amplitudes corresponding to the infinitely many alternate ways of going from A to B. In a sense, what really happens on the subatomic level is the sum of all that might have happened.

Wheeler put him to work on the problem of radiation reaction, the effect on an electron relayed to it through its own field owing to the electron's motion. It was a frontier question, and Feynman was asked to give his first technical talk on it at Princeton. He finds to his horror that in the front row sit all the local "monster minds"; Eugene Wigner, Henry Norris Russell, John von Neumann, Wolfgang Pauli, and Albert Einstein. But another important self-discovery comes to the rescue: "A miracle occurred, as it has occurred again and again in my life.... The moment I start to think about physics, nothing else occupies my mind."

This becomes his trademark. When he is recruited to the wartime program at Los Alamos, his graduate work barely finished, he, "Little Richard," is assigned to work under the gentle giant of physics, Hans Bethe. Bethe really knows pretty much everything but he needs someone to talk to: "Well, he [Bethe] comes in to this little squirt in an office and starts to argue, explaining his ideas. I say, 'No, no, you're crazy. It'll go like this,.... It turned out that's exactly what he needed. I got a notch up on account of that.'"

One gets with this story a fleeting look into the sociology of the profession. The style here was far less formal and hierarchical than in most European and Asian labs of that time. In good part through their successes, the unbuttoned Feynmans and Jim Watsons have become the reigning model of the brilliant scientist. It is a modern type whose credo is that talent is all, rank is little; you can never believe the experts, but you can and must work it out yourself; you say whatever is on your mind if you are really good enough; and you can follow your drives and instincts without shame. We recognize here a kin of Jean Jacques Rousseau's child of nature, his gifts unspoiled and uninhibited by what Feynman calls the "baloney" that pompous fools like to pass off as high civilization.

All this has a hard-edged, aggressive sound—but it is a credo directly based on a passionate view of what scientific integrity demands. In the final pages, Feynman confesses what he believes to be at the heart of scientific integrity: "A kind of utter honesty, a kind of leaning over backwards.... The first principle is that you must not fool yourself—and you are the easiest person to fool. So, you have to be very careful about that." There was an important connection between his splendid science and the sometimes outrageous behavior he reported in his autobiography. He does not try to fool *others* about himself, because thereby he is less likely to fool himself. In the service of Feynman's definition of utter honesty, it counts for little that others may now and then be scandalized: after all, it preserves the freedom of spirit for daring in his physics, too.

Nor does Feynman mean to give advice that anyone else can follow lightly. At the end of the very last page, he writes that it is only the lucky ones who "are free to maintain the kind of integrity I have described." After some 300 pages of laughter, it dawns on the reader what the personal cost was of serving this high goal, and how painful the hairshirt sometimes must have been that this unique character put on himself.

Reference

1. Richard P. Feynman as told to Ralph Leighton, *"Surely You're Joking, Mr. Feynman!"*: *Adventures of a Curious Character* (New York: Norton, 1985).

Michael Polanyi
and the History of Science

Biographical Background

I shall be discussing Michael Polanyi's career as a scientist and philosopher from the point of view of history of science as well as his contribution to the history of science. But in order to arrive at my conclusions, I shall have to make some excursions into other fields, such as the branches of politics and epistemology that were fashionable during his career. For it is fair to say that Polanyi (1891–1976) was a unique person in the history of science, not least in the spectrum of his interests. The *Dictionary of Scientific Biography* (Vol. 18) lists his professional fields as chemistry, philosophy, and sociology.

Polanyi's early family upbringing in Budapest must have counted a great deal in the eventual choice of his ideological direction. His parents' home was a place for "regularly held literary gatherings that attracted the leftist intelligentsia, some of them Marxists."[1] Moreover, the three children of the family joined left-wing youth movements at an early age. But as in the case of so many Central European intellectuals, the subsequent course of history radicalized Michael against authoritarian political movements.

He went on to study medicine at the University of Budapest, graduating a year before World War I. Even before obtaining his diploma he was publishing in chemistry, and although qualifying as a physician, he also earned a Ph.D. in chemistry with a thesis on thermodynamics. In the Albert Einstein Archives is a considerable amount of Einstein-Polanyi correspondence. It begins with a letter of 30 January 1913, from Einstein in Zurich to Professor Bredig in Karlsruhe, who had been unable to judge Polanyi's thesis on entropy at high pressures and who had sought advice on it from Einstein. The latter responded, "The publications of your Herr Polanyi please me very much." Einstein had checked them and found them altogether useful and full of fortunate thoughts. (He only wished that it had been done at less length.) Then follows a series of handwritten letters to Polanyi, in which Einstein does find some points of disagreement on matters of thermodynamics. But they clearly speak as equals. Michael's son, John C. Polanyi, later noted that Polanyi's first scientific paper was published on Einstein's recommendation, adding, "it really marked the first step in my father's academic career."[2]

Polanyi's interest in the history of science was perhaps triggered by his personal experience of historic episodes. When he presented his theory at a meeting in Berlin, at which Einstein was another participant, it was rejected and it took more than a decade until his views began to gain acceptance. Eventually, Polanyi became a researcher at the Kaiser Wilhelm Institute for Physical Chemistry in Berlin, where he developed his best-known work on dislocation theory. Leaving Germany when the Nazi party seized power in 1933, he became Professor of Physical Chemistry at the University of Manchester, where he continued a period of high achievements in physical chemistry. But perhaps under the pressure of contemporary history, his interest turned to the social sciences, and he resigned his professorship in 1948 to move to a chair in social studies. After his retirement in 1958 he continued his researches in the sociology and philosophy of science at Merton College, Oxford.

Two Sources of Polanyi's Approach

To understand the animus of his philosophical-sociological views, one must remember that throughout most of Polanyi's career as a scientist, a main model for philosophy of science for the scientific community came from the writings of logical positivists or logical empiricists, who in their most extreme writings seemed to say that science is based only on sense data and that all questions that cannot be subjected to tests in the laboratory are meaningless. But Polanyi's own work had convinced him that while objective knowledge is the basis of science as transmitted within and by the scientific community as a whole, during the nascent period of a research program or new idea, individual scientists draw on resources quite different from those that follow the logic of justification of proven achievements. Some prominent philosophers thought otherwise. Thus in a passage quoted by Polanyi, Hans Reichenbach, in the essay "The Philosophical Significance of the Theory of Relativity," had written that "the philosopher of science is not much interested in the thought processes which lead to scientific discovery; he looks for a logical analysis of the completed theory, including the relationships establishing its validity. That is, he is not interested in the context of discovery, but in the context of justification."[3] To be sure, Reichenbach at once modified this severe judgment by adding: "But the critical attitude may make a man incapable of discovery; and as long as he is successful the creative physicist may very well prefer his creed to the logic of the analytic philosopher." Yet, one should also note that no sooner had he made this generous gesture he followed it up by writing: "The philosopher has no objections to a physicist's beliefs so long as they are not advanced in the form of a philosophy."[4]

In opposition to all this, Polanyi stressed the value of what he called "tacit knowledge," the "personal notions and concepts" that are essential for progress and motivation, but that may be neither sanctioned by the scientific theories of the moment nor even consciously known to the creating scientist. I shall have much to say about this later, including a surprising test of the concept. But Polanyi went further by also opposing reductionism, the widely held conviction that ultimately all biological and social phenomena will yield to the primacy of explanation in terms of mathematics, physics, and chemistry—and one should add that since the triumph of quantum mechanics most physicists tend to think that chemistry itself is just that part of physics which really works.

A radicalizing event in Polanyi's life appears to have been his visit to the Soviet Union in the mid-1930s, where, he later wrote, he first encountered questions of philosophy. During a discussion with Bukharin in Moscow in 1935, Polanyi was especially appalled by the concept of "planning and guidance of scientific research."[5] He returned from the trip eager to devote himself to work on behalf of the freedom of scientists to choose the content, subject, and means of their work. Thereby he ran head-on into the opposition from a strong movement in the United Kingdom, led by such well-known scientists as J. D. Bernal, J. B. S. Haldane, and P. M. Blackett. The Society for the Freedom of Science, which Polanyi founded and which attracted many adherents in this country also (for example, P. W. Bridgman at Harvard), was no match for such prominent adversaries.

As it turned out, with the explosion of research opportunities in the post–World War II era, the support system in most countries in the West showed itself capable of providing scientists with enough elbow room to pursue their own ideas. But during the prewar years in Britain, the ideas expressed by scientists with Marxist leanings were reinforced by a famous conference on the history of science held in the early 1930s, in which a number of Soviet scholars made presentations. The most impressive and memorable among them was Boris Mikhailovich Hessen, who published a work entitled "The Social and Economic Roots of Newton's *Principia*." Hessen followed the agenda set forth by Friedrich Engels, whose views on the history of science were the most commanding element in Hessen's whole approach. I should inject here that when I visited China some years ago to give some lectures on the history of science under the auspices of the Chinese Academy of Sciences, I found that for most Chinese scholars the basic work in the history and philosophy of science was still thought to be the set of notes Engels had prepared for himself in the period between 1873 and 1882, published in 1927 under the title *Dialectics of Nature*. To Engels, "A single achievement of science like James Watts' steam engine has brought in more for the world in the first 50 years of its existence than the world has spent on the promotion of science since the beginning of time."[6] Science itself should thus be examined in the light of Marx's theory of historical materialism. As Engels had put it, "It is not the consciousness of men that determines their existence, but their social existence that determines their consciousness."[7] And, Engels added, "From the very beginning the origin and development of the sciences has been determined by production."[8]

In this spirit, Hessen, on British soil, attacked the very icon of abstract science, Sir Isaac Newton, and proclaimed, as he put it, "the complete coincidence of the physical thematics of the period, which arose out of the needs of economics and technique, with the main content of the *Principia*."[9]

Toward the Tacit Dimension

Opposition to such currents of thought, accepted in the 1930s more widely in the West than we may now imagine, surely determined Polanyi's motivation in large part. His visit to the Soviet Union, of which I spoke earlier, moved him in the same direction. To document his views on this matter, I can do no better than quote a part of Polanyi's introduction to *The Tacit Dimension,* which he completed in April 1966. There he bared to us his soul in these words:[10]

> I was struck [during the discussion with Bukharin] by the fact that this denial of the very existence of independent scientific thought came from a socialist theory which derived its tremendous persuasive power from its claim to scientific certainty. The scientific outlook appeared to have produced a mechanical conception of man and history in which there was no

place for science itself. This conception denied altogether any intrinsic power to thought and thus denied also any grounds for claiming freedom of thought.... My search has led me to a novel idea of human knowledge from which a harmonious view of thought and existence, rooted in the universe, seems to emerge.

I shall reconsider human knowledge by starting from the fact that *we can know more than we can tell.*

One sees in the italicized phrase the head-on attack against the instrumentalism of the positivists who, ever since Galileo made the division between primary and secondary qualities, have insisted that we can know for certain only what can be rendered in quantifiable, shareable terms. And it is also a confrontation with the Wittgensteinian positivism which, in the last sentence of the *Tractatus,* declares that whereof we cannot speak, thereof we must be silent—the implication being, of course, that such things are not worth thinking about.

Perhaps I should be permitted here a brief digression of a personal nature. I came to know and like Michael Polanyi during the last two decades of his life, when he often visited the United States. He was always gracious to younger colleagues; but we also shared a reaction against the more extreme form of positivism—although in those postwar decades, the Vienna Circle type of positivism had become greatly ameliorated as its proponents took interest in sociological, psychological, and historical components, and had moved away from the strict form that characterized its early phase. I had discovered through the study of the history of science that the model of science in terms of observable phenomena and the logic of analysis and mathematics alone by no means accounts for the creation, discussion, acceptance, rejection, and ultimate fate of any scientific advance. A clear distinction had to be made between two meanings of science: the personal stage of science, and the corpus that gets into textbooks as public science.

This distinction became particularly clear to me while assembling and studying the archives of Einstein's correspondence and manuscripts at Princeton. I found myself in need of a way of thinking about the growth of science that would add a dimension to the logical-analytic and the phenomenic dimensions. In my case I opted for including a third dimension in the study of the origins of scientific thought, the thematic one. That is not the subject of this chapter; but it indicates why I had sympathy for Polanyi's wish to escape from a two-dimensional view of science.

But to return to Polanyi's presentation of the *Tacit Dimension.* Immediately after the passages quoted above, ending with the point that "we can know more than we can tell," he adds, "This fact seems obvious enough; but it is not easy to say exactly what it means." And he gives the example of pattern recognition—for example, the recognition of somebody's face. Polanyi tries to make it plausible that one can know and act on what one cannot tell, by referring to Gestalt psychology. He refuses to think that Gestalt psychology can be reduced to the disposition of impressions on the retina or the brain. Rather, to him Gestalt is "the outcome of an active shaping of experience performed in the pursuit of knowledge. The shaping or integrating I hold to be the great and indispensable tacit power by which all knowledge is discovered, and once discovered is held to be true."[11] And it is not only perception which is an instance of tacit knowing. More generally, "our bodily processes participate in our perceptions";[12] and once we understand that this is the case we can "throw light on the bodily roots of all thought, including man's highest creative powers."[13]

This argument is analogous to Wilhelm Dilthey's teaching that an empathy or in-dwelling is required for a proper knowledge of man and the humanities, an aesthetic appreciation that may not be possible to render directly in language. Polanyi believes the same to be the case for the natural sciences also. "To rely on a theory for understanding nature is to interiorize it. For we are attending from the theory to things seen in its light and are aware of the theory, while thus using it, in terms of the spectacle that it serves to explain. This is why mathematical theory can be learned only by practicing its application: its true knowledge lies in our ability to use it."[14] Formalizing all knowledge, "to the exclusion of any tacit knowing," which is the aim of those who hold to the model of "strictly detached objective knowledge," is "self-defeating, for in order that we may formalize the relations that constitute a comprehensive entity, for example the relations that constitute a frog, this entity, i.e., the frog, must be first identified informally by tacit knowing."[15] And Polanyi adds that the most "striking concrete example of an experience that cannot possibly be represented by any exact theory" is simply "the experience of seeing a problem, as a scientist sees it in his pursuit of discovery."[16] What you need most at that stage is "the intimation of something hidden, which we may yet discover." It is this "tacit foreknowledge of yet undiscovered things"[17] which provides the sometimes passionate motivation to uphold a direction of work or a theory against heavy pressures from the outside.

Elaboration and Responses

Polanyi's approach has two results. One is that he is in head-on conflict with Karl Popper, by severely disagreeing with Popper's doctrine that "the scientist is not only indifferent to the outcome of his surmises, but actually seeks their refutation."[18] Polanyi responds: "This is not only contrary to experience but logically inconceivable. The surmises of a working scientist are born of the imagination-seeking discovery. Such effort risks defeat, but never seeks it. It is in fact his craving for success that makes the scientist take the risk of failure. There is no other way."[19] Obviously, a working scientist is speaking here.

The second result is that Polanyi touches here, without saying so, on the very old problem of how major discovery is possible in the first place, given the obvious, severe limitations of the human mind faced with the infinitude of natural phenomena and their connections. Einstein offered a solution to this problem in 1918 with the daring suggestion that our minds are guided by "what Leibniz termed happily 'the preestablished harmony.'"[20] You will recall that Gottfried Wilhelm Leibniz had postulated that our ability to discover the laws concerning material bodies is one aspect of the unity from which God created the two apparently separate entities of the universe, the spiritual and the material. Each of these obeys its own laws, but each can interact in sympathetic unison, somewhat in the way one string instrument goes into resonance and picks up the sounds made by a second one that is tuned to it. Or to use Leibniz's own words to explain this possibility of harmonious interaction, in which he uses an image that must have delighted Einstein, "the souls follow their laws and the bodies follow theirs, but nevertheless these two beings of entirely different kind meet together and correspond to each other like two clocks perfectly regulated at the same time. It is this that I call the theory of preestablished harmony."

In the early nineteenth century, the Danish physicist Hans Christian Oersted also struggled with this problem; in his way of reading Immanuel Kant, which was typical for the *Naturphilosophen,* Kant's insistence that phenomenal facts are not things in themselves but mere appearances culminated in the warning that the study of these appearances and the connections between them are an interaction not with nature but with one's own mind. As Kant had put it in the *Critique of Pure Reason:* "That nature should direct itself according to our subjective ground of apperception, and should indeed depend upon it in respect of its conformity to law, sounds very strange and absurd. But when we consider that this nature is not a thing in itself but is merely an aggregate of appearances, so many representations of the mind, we shall not be surprised that we can discover it only in the radical faculty of all our knowledge, namely in transcendental apperception, in that unity on account of which alone it can be entitled the object of all possible experience—that is, nature."[21]

Oersted, and I think Polanyi also, found this idealism uncongenial to a working scientist's mind, and Oersted therefore invented a modification, in his splendid conception of an "anticipating consonance"[22] existing between the mind of the scientist and the workings of nature. Polanyi comes very close to this notion. For example, he writes: "When a discovery solves a problem it is itself fraught with further intimations of an indeterminate range, and...when we accept the discovery as true, we commit ourselves to a belief in all these as yet undisclosed, perhaps as yet unthinkable, consequences. This is of course not explicit knowledge, and [he acknowledges] there is no explicit justification for the perception of a dawning truth."[23]

Still, Polanyi never quite admitted that these elements of tacit knowledge and of intimations of undisclosed consequences are, more often than not, simply wrong, although as a working scientist he must have observed this to be the case. But he would also have known that there is a certain pattern to these intimations, or anticipations. Almost by definition, a major scientist is one for whom this mechanism somehow works, at least often enough. At any rate, whether these anticipations are correct or turn out to be (in his words), "a delusion."[24] Polanyi holds it futile to search for strictly impersonal criteria of their validity, "as positivistic philosophies of science have been trying to do for the past 80 years or so."[25]

In the climate of the decay of logical positivism after the mid-1950s, Polanyi's conception of tacit knowledge, or personal knowledge, did not remain a prominent target of attack. In any case it was never really a completely thought-through theory of scientific creativity. On the other hand, a concept which Polanyi held to be directly related to the tacit dimension, namely "emergence," seems to me to have been then, and remains now, a focus of debate and opposition, particularly among biologists. To rescue biology from reductionism, from being dissolved into mere physics and chemistry, Polanyi announced "the principle that the operations of a higher level can never be derived from the laws governing its isolated particulars,"[26] hence that "none of the biotic operations can be accounted for by the laws of physics and chemistry."[27]

Here we encounter a newer version of the old debate which so agitated scientists and philosophers in the nineteenth century, of mechanism versus vitalism. Of course Polanyi did not deny that there is "a great deal of truth in the mechanical explanation of life";[28] but he wanted to insist that living functions are "determined at all stages by a combination of a mechanism with organismic regulation."[29] At the very least, he said, "a principle not present in the inanimate must come into operation when it gives birth to living things."[30] Such views, coming from a prominent physical chemist, found probably a much more willing audience outside the laboratory than in it.

Also, to Polanyi, the principal interest of evolution was the rise of man from "lower" beings. To him, the problem of evolution seems to boil down to understanding how we reached "our position as the highest form of life on earth, and our own advent by a process of evolution."[31] But we are now in the age of anti-speciesism, in which even the Spotted Owl has still some political clout; so one cannot expect much resonance nowadays with Polanyi's call for a reshaping of "the problem of evolution deformed by the current theory of evolution."[32] The origin of species was a preoccupation which he thought can only make us "lose sight" of that more fundamental question. Properly understood, evolution is an expression of the concept of the "stratified universe of living things,"[33] in which progress from one level to the other cannot be made via reduction, or even by the continuation of the logic of one level with respect to the logic of the second above it, but rather by emergence—"the first emergence by which life comes into existence being the prototype of all subsequent stages of evolution."[34] Polanyi is quite frank that such ideas connect with earlier versions encountered in the history of science, for example, that of Teilhard de Chardin.[35]

With such tools, Polanyi struggled with what he called the concept of the "potentiality for obedience to higher demands"[36] and "the capacity to feel reverence for men greater than oneself,"[37] both of which he regarded as aspects of the process of evolution. One must acknowledge that within a system of concepts very different from Polanyi's, sociobiologists such as E. O. Wilson are in fact struggling with very similar problems in their discussion of altruism. I mention this only to indicate what to an historian of science is again and again so impressive: the continuity of preoccupations of the same sort within very different frameworks and worldviews.

Consequences of Polanyi's Doctrines

As if by simple extrapolation, we can guess where Polanyi's thought might land next. It is the modern base for moral belief. How, he asked, can "intellectual powers, grounded in tacit knowing and descended from evolutionary emergence...exercise the kind of responsible judgment which we must claim if we are to attribute a moral sense to man. In a world where, it is widely held, scientific rationalism has impaired moral beliefs...by shattering their religious connections, where the Enlightenment weakened ecclesiastical authority, and modern positivism denied justification to all transcendent values,"[38] where can one find a theory for reestablishing the justification of moral standards? Control through established ecclesiastical authority appealed to Michael Polanyi as little as the control of science itself. Thus he wrote, "It was only when the philosophy of Enlightenment had weakened the intellectual authority of the Christian churches that Christian aspirations spilled over into man's secular thoughts and vastly intensified our moral demands on society. The shattering of ecclesiastical control may have been morally damaging in the long run, but its early effect was to raise the standards of social morality."[39]

What he feared most of all was the fusion of scientific skepticism and moral perfectionism. He saw that hybrid represented by modern existentialism and by what he called "an angry absolute individualism."[40] And the same hybrid also, by demanding a total transformation of society as a utopian project, expressed itself in Marxism as a political doctrine. In fact, it would be difficult to associate Polanyi's ideas with allegiance to any "ism"; for by giving first priority to the concept of tacit thought as an indispensable element of all knowing, by claiming that "the transmission of knowledge from one generation to the other must be predominantly tacit,"[41] Polanyi therefore denied the possiblity that any body of knowledge could be concretized in a uniquely shareable ideology at a given time or through history.

And yet, apparently paradoxically, Polanyi sees a way of attempting total individual understanding with one's own mental faculties. That alternative is "entrusting oneself...to a teacher or leader."[42] St. Augustine made a similar observation when he taught, "Unless you believe, you shall not understand." However, Polanyi does not hold out great hope that religion as now understood could fill this place for the need for tradition. "Modern man's critical incisiveness must be reconciled with his unlimited moral demands first of all on secular grounds. The enfeebled authority of revealed religion [as he called it] cannot achieve this reconciliation; it may rather hope to be *revived* by its achievement."[43]

Polanyi's final paragraph in *The Tacit Dimension* indicates his puzzled frame of mind on this point: "Perhaps this problem cannot be resolved on secular grounds alone. But its religious solution should become more feasible once religious faith is released from pressure by an absurd vision of the universe, and so there will open up instead a meaningful world which could resound to religion."[44] This view is connected, I believe, with Einstein's much better developed ideas on Cosmic Religion, with which Polanyi was undoubtedly acquainted.

I don't see it as my task to provide a rebuttal to, or even a general assessment of, Polanyi's thoughts. This has been done many times, for example, in the volume *Intellect and Hope,* edited by Thomas A. Langford and William H. Poteat. But perhaps a few words are appropriate about the way comments on *The Tacit Dimension* generally run. Let me refer here to Robert S. Cohen's essay in the volume edited by Marjorie Grene, *Interpretations of Life and Mind,* an essay entitled "Tacit, Social and Hopeful." Cohen, both a physicist and a philosopher of science, acknowledges right away that the tacit dimension of knowing appears to him "acceptable and well-established."[45] Polanyi did not discover tacit knowledge, but he discovered at least how important it was in his own epistemology, and made more of it than many others. For Polanyi, Cohen writes, "knowledge is situated within a background of clues, or a tacit background." That "means that there is a reality hidden behind the discovered objects. And so objects as we know them become clues to an as yet undiscovered and deeper level of reality."[46] For philosophy, this poses an immense challenge owing to the whole series of progressively more hidden realities it implies.

But, Cohen asserts, there is in all this a hint that Polanyi has a novel and interesting though undeveloped view of how the history of science progresses. "Polanyi asserts that different epochs of science offer different cue-maps, different forms of in-dwelling."[47] Any working scientist who has passed through the development of his or her own field over a couple of decades (and nowadays that can be an immense distance)—with the possibility of vast changes of mind and attitudes—is likely to assent to this picture.

The "Big Book" and "the Story of Relativity"

The work that most closely connects Polanyi with the field of the history of science as scholarship is what he called his "big book," *Personal Knowledge: Toward a Post-Critical Philosophy,* published in 1958. As he says at the beginning of his preface,

> This is primarily an inquiry into the nature and justification of scientific knowledge. But my reconsideration of scientific knowledge leads on to a wide range of questions outside science. I start by rejecting the ideal of scientific detachment. In the exact sciences this false ideal is perhaps harmless, for it is in fact disregarded there by scientists. But we shall see that it exercises a destructive influence in biology, psychology and sociology, and falsifies our whole outlook far beyond the domain of science. I want to establish an alternative ideal of knowledge, quite generally. Hence the wide scope of this book and hence also the coining of the new term I have used

for my title: personal knowledge. The two words may seem to contradict each other: for true knowledge is deemed impersonal, universally established, objective. But the seeming contradiction is resolved by modifying the conception of knowing.[48]

That new conception is based on the view that the personal participation of the knower in acts of understanding does not make such understanding *subjective*. The act of comprehension is "a responsible act claiming universal validity. Such knowing is indeed *objective* in the sense of establishing contact with a hidden reality."[49]

The book was based on Polanyi's Gifford Lectures, delivered in 1951–52 at the University of Aberdeen. But he confesses that he spent "nine years almost exclusively on the preparation of this book."[50] Nevertheless, as even his friends and followers admit, it is by no means an easy book to read or accept. In their introduction to the collection, *Intellect and Hope: Essays in the Thought of Michael Polanyi*, the editors, Thomas A. Langford and William H. Poteat, begin with the sentence: "*Personal Knowledge* is an exasperating book." They add, "If one does not find it exasperating, one has not *really* read it…. There can be no doubt that *Personal Knowledge* comes to us with its rhetoric all out of focus. It is a mixed bag…. Philosophers by and large, at least English-speaking philosophers on both sides of the Atlantic, find *Personal Knowledge* annoying because it is dangerously loose, innocuous because it says what has been said elsewhere and better, or irrelevant because its preoccupations are no legitimate concern of philosophy or of philosophers…. One never 'gets going.'"[51]

There is no doubt that the book is maddening in spots. For example, Polanyi writes, "The principal purpose of this book is to achieve a frame of mind in which I may hold firmly to what I believe to be true, even though I know that it might conceivably be false."[52] It has been suggested that it may be best to consider it an example of the confession literature, with such distinguished antecedents as Augustine and Rousseau.

Polanyi begins by going over a version of the development of the ideas of the solar system during the scientific revolution of the seventeenth century as a consequence of the Copernican model, and he draws on various well-known anecdotes to show how foolish it would be to hold that these early scientists forbade themselves to "go beyond experience by affirming anything that cannot be tested by experience."[53] Quotations of ecstatic passages from Kepler easily show the opposite to be the case. But Polanyi devotes only a few pages to these matters, for his main proof depends on what he calls "the story of Relativity." That theory was indeed taken by the positivists to show that through instrumentalist thinking Einstein had freed nineteenth-century physics from its metaphysical underpinnings, and thereby made the breakthrough to modern science. Polanyi correctly points out that every textbook of physics tried to present the rise of relativity as the necessary response to an experimental situation, namely the supposed null result of the Michelson-Morley experiment searching for an ether drift in 1887—fully in accord with the sensationist or positivist view of how theories must proceed. (As well, we should add, as being the easiest *pedagogic* method of convincing students that they must take seriously what otherwise would be so counterintuitive.) But, Polanyi declares, "the historical facts are different."[54] Noting that Einstein, in his publication, had not mentioned the Michelson-Morley experiment at all, he concludes that this theory was proposed "on the basis of pure speculation, rationally intuited by Einstein before he had ever heard about it."[55]

An Experimental Proof of Tacit Knowledge

Let us stop at this important point in Polanyi's book and consider what an historian of science would now do on the basis of such a personal hunch or presupposition. It is an interesting enough case to give it serious treatment. You would begin by searching the literature of the period around the publication of the theory, encompassing perhaps a decade to either side of it. You would read not only Einstein but also his contemporaries, to see who says what, if anything, about the Michelson-Morley experiment and other experiments of the same sort by 1905. Then you would try to consult available documents in the archives of the main persons involved in the genesis and debates, pro and con, on the special theory of relativity, in the hope of finding contemporaneous exchanges or unpublished drafts and manuscripts from Einstein, but also from H. A. Lorentz, H. Poincaré, etc. You would also try to consult oral history interviews, autobiographical writings at a later stage, and so forth. If a promising Ph.D. candidate had come to me with such a project, I would have estimated it would take a year or two of research and quite a bit of travel to archives. In fact, I published a long article on this case in *Isis* in 1969, entitled "Einstein, Michelson, and the 'Crucial Experiment,'" and it did take me the better part of a year.

But Michael Polanyi did none of that. As he tells us in his book *Personal Knowledge,* he availed himself of a remarkable shortcut. After all, he had been in touch with Einstein since 1913. Taking advantage of his entrée, and in order "to make sure" of his hunch that Einstein's theory was based on "pure speculation, rationally intuited by Einstein before he had ever heard" of the Michelson-Morley experiment, Polanyi got in touch with the physicist N. Balazs, who was working with Einstein in Princeton in the summer of 1953, and asked his fellow former-countryman to submit this speculation to Einstein himself.

As a consequence, Balazs had an interview with Einstein on that subject, describing it to Polanyi in a letter of 8 July 1953. He reported that Einstein concurred that (as Balazs wrote) "The Michelson-Morley experiment had no role in the foundation of the theory. He got acquainted with it while reading Lorentz's paper about the theory of this experiment (he of course does not remember exactly when, though prior to his papers), but it had no further influence on Einstein's considerations, and the theory of relativity was not founded to explain its outcome at all."[56] What did matter during the genesis of his ideas, Einstein had told Balazs, was his concern with a series of more fundamental problems, such as the impression an observer, moving with the velocity of light, would have while viewing the light wave, and the lack of symmetry of action between coils and magnets when they are moved with respect to each other in producing the induction of currents in the coil.

Needless to say, this second-hand report of what Einstein may have said to Balazs, which Polanyi strangely chose to quote only in a footnote in his book, was not found convincing either by philosophers of science or by historians of science, the more so as the book as a whole was using this report as a tool in an otherwise quite idiosyncratic attempt at a new epistemology. Not until years later, when all the supporting work that I have mentioned above as necessary was actually done, would there be the kind of impact on the scholarly community which Polanyi had hoped to make by his shortcut.

And yet, and yet...Polanyi *was* right. His hunch, of which he was so convinced that he tested it only in the most perfunctory way, through a third party (not even taking the trouble to put the question to Einstein himself), was borne out later by a great deal of more laborious work. To be sure, Polanyi overreached when he declared that Einstein's theory was framed "on the basis of pure speculation, rationally intuited"; but the theory was evidently based chiefly both on the speculation about a thought experiment (that of traveling with the speed of light along a light beam) and some old, well-established nineteenth-century experiments long before Michelson's, those of Faraday, Fresnel, and of stellar aberration.[57]

How could that be? Was it merely an accident that Polanyi's presupposition was, on the whole, borne out? Perhaps. But I prefer to think of it in Polanyi's own terms. After all, for decades he had been a very prominent and successful scientist himself, engaged both in experiment and theory. He had *internalized* how scientists think; he had observed how others do their work, in finished publications as well as in conversations and in debates—for example, during his time in Berlin, when Einstein was also there and Polanyi saw much of him. In short, if there is such a thing as apperception, personal or tacit knowledge, and in-dwelling, we must allow Polanyi to have had those capabilities as a scientist himself. Or to put it in Hans Christian Oersted's terms: Polanyi's prediction of how serious research in the entirely different profession of the history of science would illuminate the genesis of relativity was an act of anticipating consonance with the real state of affairs, one that would be made fully clear only later.

In short, I would like to hold open the possibility that it is precisely Polanyi's *lack* of having made a serious study and yet having reached the right sort of conclusion that constitutes, as it were, an experimental verification of his concept of personal and tacit knowledge. We all know that this sort of mechanism has worked in science from the days of Kepler and Galileo, who made advances to which their purely scientific knowledge of the time did not really entitle them. Polanyi is the first example I know of where the same sort of thing happened in the pursuit of the history of science.[58]

References

1. Ferenc Szabadvaary, *The Dictionary of Scientific Biography (DSB)* (New York: Scribner's, 1975), vol XVIII, p. 718.
2. Letter of 5 June 1964 from J.C. Polanyi to Helen Dukas (who had been Einstein's secretary).
3. H. Reichenbach, in *Einstein: Philosopher-Scientist*, ed. Paul A. Schilpp (Evanston, Il.: Library of Living Philosophers, 1949), p. 289. This reference is given by Polanyi in *Personal Knowledge* (Chicago: U. of Chicago Press, 1958), p. 14, footnote 1 (along with a quote from H. Mehlberg expressing a similar point of view).
4. Ibid. Similar passages can be found in many places, for example in Karl Popper's *The Logic of Scientific Discovery* (New York: Harper and Row, 1959), p. 31.
5. See Ferenc Szabadvaary's summary of Polanyi's views in *DSB*, p. 719: "In the mid 1930's, Polanyi visited the Soviet Union. After his return, he wrote several articles and a book criticizing Soviet economic notions, and on planning and guidance of scientific research....He was a supporter of complete freedom in scientific research, not only concerning its content but also in the choice of the subject."
6. F. Engels, "Outlines of A Critique of Political Economy," in *Collected Works* (New York: International Publishing Co., 1975), vol II, p. 428. Quoted and referenced by Robert S. Cohen in *DSB*, vol. XV, p. 136.
7. F. Engels, *Preface to A Contribution to the Critique of Political Economy* (New York: 1970), pp. 20–21. Quoted by R.S. Cohen in *DSB*, vol. XV, p. 136.
8. F. Engels, *Dialectics of Nature* (Moscow: 1954), p. 247. Quoted by R.S. Cohen in *DSB*, vol. XV, p. 136.

9. B. M. Hessen, *The Social and Economic Roots of Newton's Principia*, introduction by Robert S. Cohen (New York: H. Fertig, 1971), p. 26.
10. M. Polanyi, *The Tacit Dimension* (New York: Anchor Book A540, Doubleday, 1967), pp. 3–4. The concept of tacit knowledge in science has also been treated by others, though not necessarily with the same terminology. For a recent example, see Gerhard Grössing, *Das Unbewusste in der Physik* (Vienna: Verlag Turia & Kant, 1993).
11. Ibid., p. 6.
12. Ibid., p. 7: "...bodily processes are prominent in the operations of perception." Similarly, Einstein reports that what he called "the elements of thought" were "in any case of visual and some muscular type." [Jacques Hadamard, *The Psychology of Invention in the Mathematical Field* (Princeton, N.J.: Princeton University Press, 1945), pp. 142–143.] See also Edmund Husserl, who writes: "Moving freely within the moment of experience which brings what is present into my intuitional grasp, I can follow up these connections of the reality which immediately surrounds me. I can shift my standpoint in space and time, look this way and that, turn myself forwards and backwards; I can provide for myself constantly new and more or less clear and meaningful perceptions and representations, and images also more or less clear, in which I make intuitable to myself whatever can possibly exist." [Edmund Husserl, *Ideas*, trans. W. R. Boyce Gibson (New York: Collier Books, 1962), chapter 3, section 27, p. 92.] See, further, Maurice Merleau-Ponty: "We grasp external space through our bodily situation. A 'corporeal or postural schema' gives us at every moment a global, practical, and implicit notion of the relation between our body and things....A system of possible movements, or 'motor projects,' radiates from us to our environment. Our body is not in space like things; it inhabits or haunts space. It applies itself to space like a hand to an instrument...." ["An Unpublished Text by Maurice Merleau-Ponty," trans. Arleen B. Dallery, in Maurice Merleau-Ponty, *The Primacy of Perception*, ed., James M. Edie (Evanston, Il.: Northwestern University Press, 1964), p. 5; first published as "Un inedit de Maurice Merleau-Ponty," *Revue de metaphysique et de morale*, no. 4 (1962), 401–409.]
13. M. Polanyi, *The Tacit Dimension*, p. 15.
14. Ibid., p. 17.
15. Ibid., p. 20.
16. Ibid., p. 21.
17. Ibid., p. 21: "For to see a problem is to see something that is hidden. It is to have an intimation of the coherence of hitherto not comprehended particulars." Again, compare with Husserl: "What is actually perceived ...is partly pervaded, partly girt about with a *dimly apprehended depth or fringe of indeterminate reality*....Moreover, the zone of indeterminacy is infinite. The misty horizon that can never be outlined remains necessarily there." [E. Husserl, *Ideas*, p. 92.] See also Merleau-Ponty: "The characteristic property of the Visible is to have a layer of invisibility in the strict sense, which it makes present as a certain absence." ["Eye and Mind," trans. Carleton Dallery, in M. Merleau-Ponty, *The Primacy of Perception*, p. 187; first published as "L'Oeil et l'Esprit" in *Art de France*, vol. 1, no. 1 (January, 1961).] For a discussion, see E. Wigner and R. A. Hodgkin, *Bibliographical Memoirs of Fellows of the Royal Society*, vol. 23 (1977), p. 430, footnote.
18. Polanyi, *The Tacit Dimension*, p. 78. Polanyi quotes Popper's words and gives the reference in footnote 10, p. 98: "This view has been persuasively expressed by K. R. Popper, e.g., in *The Logic of Scientific Discovery*, New York, 1959, p. 279."
19. Ibid., pp. 78–79.
20. A. Einstein, in an address given at a celebration of Max Planck's sixtieth birthday before the Physical Society in Berlin. In *Ideas and Opinions* (New York: Bonanza Books, 1954), p. 226.
21. I. Kant, *Critique of Pure Reason*, A 114.
22. H. C. Oersted, "On the Spirit and Study of Universal Natural Philosophy" in *The Soul of Nature*, trans. Leonora and Joanna B. Horner (London, 1852; reprint London, 1966), p. 450.

23. Polanyi, *The Tacit Dimension*, p. 23.
24. Ibid., p. 25: "The anticipation of discovery, like discovery itself, may turn out to be a delusion."
25. Ibid., p. 25.
26. Ibid., p. 36.
27. Ibid., p. 42: "Moreover, the conclusion that machines are defined by the fact that boundary conditions expressly left open by physics and chemistry are controlled by principles foreign to physics and chemistry, makes it clear that it is in respect of its characteristic boundary conditions that a mechanically functioning part of life is not explicable in terms of physics and chemistry."
28. Ibid., p. 42, immediately following the preceding quote.
29. Ibid., p. 43.
30. Ibid., p. 44. (The very definition of emergence.)
31. Ibid., p. 47.
32. Ibid., p. 46.
33. Ibid., p. 50.
34. Ibid., p. 49.
35. See ibid., p. 46. See also *Personal Knowledge*, p. 388, where he associates his views with "noogenesis" and cites Teilhard in a footnote.
36. Ibid., p. 52.
37. Ibid., p. 52.
38. Ibid., p. 56.
39. Ibid., p. 57.
40. Ibid., p. 59.
41. Ibid., p. 61.
42. Ibid., p. 61.
43. Ibid., p. 62.
44. Ibid., p. 92.
45. R. S. Cohen, "Tacit, Social and Helpful," in *Interpretations of Life and Mind,* ed. M. Grene, (London: Routledge and Kegan Paul, 1971), p. 137.
46. Ibid., p. 138.
47. Ibid., p. 140.
48. M. Polanyi, *Personal Knowledge*, preface, p. vii. I thank Mary Jo Nye for the observation that Polanyi's "tacit knowledge" and "personal knowledge" are two notions that only partially overlap.
49. Ibid., p. vii.
50. Ibid., p. ix.
51. T. Langford and W. Poteat, eds., *Intellect and Hope, Essays in the Thought of Michael Polanyi* (Durham, N.C.: Duke University Press, 1968), p. 3 and pp. 6–7.
52. M. Polanyi, *Personal Knowledge*, p. 214. T. Langford and W. Poteat quote this passage in their introduction to *Intellect and Hope*, p. 14, in the context of comparing *Personal Knowledge* to the *Confessions* of St. Augustine.
53. The import of such passages is likely to be Polanyi's opposition to the teachings of scientist-philosophers such as Ernst Mach, of whom Polanyi says (p. 9) that "his book, *Die Mechanik*, published in 1883, founded the Vienna school of positivism."
54. Ibid., p. 10.
55. Ibid., p. 10.
56. Ibid., footnote 2. Einstein's own writings name Lorentz's *book*.
57. I have discussed the relation between the Michelson-Morley experiment and Einstein's work in chapter 8 and on pp. 477–480 of G. Holton, *Thematic Origins of Scientific Thought: Kepler to Einstein* (Cambridge, MA: Harvard University Press, 1988).
58. I wish to record my thanks to Anne Davenport for helping me convert the manuscript for a lecture into an article.

Selected Bibliography

Cohen, Robert S. "Tacit, Social and Hopeful," in *Interpretations of Life and Mind. Essays around the Problem of Reduction,* ed. Marjorie Grene, (London: Routledge & Kegan Paul, 1971), 137–148.

Gelwick, Richard. *The Way of Discovery. An Introduction to the Thought of Michael Polanyi* (New York: Oxford University Press, 1977).

Holton, Gerald. "Einstein, Michelson and the 'Crucial' Experiment," in *Thematic Origins of Scientific Thought: Kepler to Einstein* (Cambridge, MA: Harvard University Press, 1988), chapter 8 and pp. 477–480.

Langford, Thomas A. and William H. Poteat, eds. *Intellect and Hope. Essays in the Thought of Michael Polani* (Durham, NC: Duke University Press, 1968).

Polanyi, Michael. "The Growth of Science in Society," *Minerva,* vol. 5, no. 4 (Summer 1967), 533–546.

Polanyi, Michael. "Life's Irreducible Structure," *Science,* vol. 160 (21 June 1968), 1308–1312.

Polanyi, Michael. *Personal Knowledge. Towards a Post-Critical Philosophy* (Chicago, IL: University of Chicago Press, 1958).

Polanyi, Michael. "The Potential Theory of Adsorption," *Science,* vol. 141 (13 September 1963), 1010–1013.

Polanyi, Michael. *Science, Faith and Society* (London: Oxford University Press, 1946).

Polanyi, Michael. *The Tacit Dimension* (Garden City, NY: Doubleday & Co., Inc., 1967).

Szabadvaary, Ferenc. "Mihaly (Michael) Polanyi," *Dictionary of Scientific Biography,* vol. 18 (New York, NY: Charles Scribner's Sons, 1981), 718–719.

Wigner, Eugene and R. A. Hodgkin. Biographical entry for Michael Polanyi, in *Biographical Memoirs of Fellows of the Royal Society,* vol.23 (1977).

Part Four:
SCIENCE FOR STUDENTS

How Can Science Courses Use the History of Science?

How can one best use the history of science in teaching a course in one of the specialty sciences? To maximize whatever useful information I might be able to present in a brief account, I will concentrate on some quite specific and personal experiences and experiments.

To set the stage, let me remind you of two kinds of paradox. One concerns the substantial interest of the larger public in the history of science, for example, in scientific biographies. Just in the field of physics, in recent years, we have had biographies or autobiographies of Bohr, Feynman, Heisenberg, Casimir, Rabi, Schrödinger, Szilard, Teller, Weisskopf, and Wigner, not to speak of the love letters of Albert and Mileva Einstein. But except for occasional anecdotes there is no spillover of this interest into recent science texts, with only a few rare exceptions. Most instructors find that their courses are ever more crowded with absolutely necessary subject matter, and they have neither the incentive nor the time to learn about the history of science sufficiently. Some historians of science who are purists do not extend a helpful hand either to the pedagogic use of the history of science outside their own courses, although there exist for this purpose a few helpful guides for the perplexed science instructor, as well as older introductory college physics texts, such as those by E. C. Kemble and by myself.[1] So despite the predisposition of many of our students for the humanistic approach, what they usually encounter in the mandatory courses for nonscientists is at best a watered-down version of the specialty course designed for the concentrator or major in a science.

The second paradox is that there is wide agreement that such courses don't really work very well for that audience; yet we continue to teach them, although they leave both students and teachers unsatisfied. The situ-

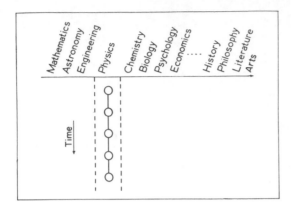

Figure 1. Traditional pres-entation of topics in intro-ductory physics.

ation is quite similar to the one in the famous story C. N. Yang told, which expresses the disenchantment of some physicists with mathematicians. When a physicist tries to get some help from a mathematician, Professor Yang said, he is usually in the position of the man who carries a large bundle of dirty clothes and searches for a laundry without success for a long time. He is greatly relieved when he finally finds a shop displaying a sign "Laundry done here" in the window. He goes in and dumps the bundle on the counter. The man behind the counter says, "What's this?" "I want to have these laundered." "We don't do laundry here." "But you have a sign in the window advertising that you do laundry." "Oh, that! We only make signs." We, too, all too often are only making signs when we teach our nonscience students.

The first of the experiments I want to describe, one which makes me feel that we have gone far beyond making only signs, is a Harvard College course I gave for many years, chiefly to beginning students, who have no commitment to science but who are obliged to spend a semester or a year in a so-called General Education or Core Curriculum course centering on a science. The mode and the purpose of these courses are quite different from those of the typical department science specialty course. As traditionally given, the introductory physics course, for example, follows roughly the topics developed from mechanics through particle physics, with very little attention to any neighboring fields. If we think of the possible universe of all courses a student might encounter in college displayed on a two-dimensional map (Figure 1), the departments offering specialty courses would range from the most mathematical on the left end of the horizontal axis to the least mathematical, say, history, philosophy, and literature, on the right end. The specialty physics course can therefore be understood as a series of blocks or a string of pearls arranged

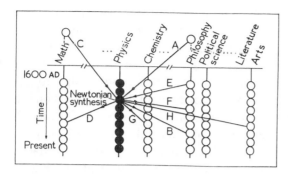

Figure 2. A specific advance in physics is linked to earlier and later achievements, not only in physics but in other fields as well.

along the vertical axis, a dimension which roughly corresponds to the time of development, from about 1600 A.D. to today. A similar string of pearls would characterize the other courses in this plane, each department jealously guarding its autonomy.

Whether this approach to education is really sound even for the future specialist is not a question I want to go into now. But it is clearly uninviting for most nonscience students. Therefore, I have designed and taught what I call a "Connective Science" course for the General Education or Core track. To put it very briefly, in addition to giving a sound presentation in physics itself, one injects, for a total of perhaps ten or fifteen percent of the time of the course, excursions into studies of certain adjoining areas which have historical relevance. This is roughly illustrated (in Figure 2) for the portion of that course when we take up Newtonian dynamics; similar diagrams will apply to other parts of the Connective Science physics course. Newton's ideas about the nature of space and time are indebted to his studies in theology and philosophy; attention to these ideas given in this course is shown by link A. Within physics itself, Newton's achievement can be seen as a culmination of the work of Kepler and Galileo, and shows up later again and again. The mathematics established by Newton's time was indebted to the Greek mathematicians (link C), and Newton repaid this debt by advancing the progress of the mathematics itself (link D). Furthermore, Newton's work affected the Deist theologians (link F) in the next century, as well as John Dalton's atomic models in chemistry in the early nineteenth century (link G), even if by a creative misunderstanding on Dalton's part. The successes and metaphors emanating from Newton's work influenced the artistic sensibilities of the eighteenth century (link H). And so forth.

The same kind of web extends from each of the other chief topics of physics to show the interactions of that science with biology, engineering, and indeed politics. Thus the intellectual map I like the student to glimpse is not a set of separate strings of beads but rather a tapestry, a fabric of ideas. Obviously a Core course or General Education course in a science cannot do full justice to this conception; but a little does go a long way to stress the relevance of science as a living part of the historic development of a whole culture. In a certain sense a view of this sort, applied to all the main areas of college study, results in what we might call education rather than mere training, an education without which the course of modern history would be almost incomprehensible. Indeed, one would hope that when students encounter the work of, say, John Locke or James Madison, Voltaire or Alexander Pope in courses in other departments, such as philosophy, politics, and literature, they would encounter, so to speak, the other end of these link-lines I have drawn and see once more the connection with their science course.

Similarly, if their science course has given them some indication of the social effects of the steam engine, motors and generators, and the nuclear reactor, they might be more sophisticated about the whole topic of science, technology, and society—which, to students of the sort I am describing now, counts for just as much as a thorough training in all the subtler aspects of $F=ma$. In short, science should be seen as an interacting element in the total intellectual activity of an age.

Obviously, teaching a Connective Science course from scratch might seem an overwhelming task. One can begin with a few such illustrations, however, and in time gently reorient one's course until there emerges this constellational pattern in place of the string-of-beads model.

I built into the course a requirement that each student write an essay on a topic of his or her choice, written after they had studied the block on Newtonian dynamics in the early part of the year. The overall title of the assignment was "The Impact of the Newtonian Synthesis on X," where X stands for the concentration or major or area of nearest interest of the student, whether it be political science or philosophy or law or psychology or fine arts, etc. In this way the student sees the course as something specific for him or her as an individual, not as an external object to be accepted on its own unyielding terms.[2]

My second example involves another audience, namely, secondary school students. In order to find a way to attract students who otherwise would not choose one of the more conventional string-of-beads type of secondary-school physics courses, James Rutherford, Fletcher Watson, and I accepted the challenge of the National Science Foundation in the early 1960s to design a physical science course with the humanistic base which I just illustrated in the previous case. Building on my experience with the General Education course described above, and having published a textbook that could serve as a general road map, we assembled a large number of collaborators and, after several trial editions and wide-ranging evaluation tests, published the result under the title *The Project Physics Course*. It is still available through Holt, Rinehart and Winston, now a subsidiary of Harcourt Brace.

A key portion of the whole process of curriculum development is, of course, the training of teachers to handle a new course. For some years Washington was serious about providing money for teacher training, so that at the high point about 250,000 students in schools (and also some in colleges) in the United States were using the main text of Project Physics and at least a substantial part of the rest of the course materials, such as readers, film loops, etc. Adaptations have been prepared in Australia, China, Japan, Portugal, Canada (one in English and one in French), Italy, and other countries. The main point of interest here is that the connective physical science aspect was embedded in *The Project Physics Course* and has been responsible for its unique structure.

Finally, let me turn to a very different problem, namely the use of history of science to help in the education of students already committed to a departmental science curriculum. A course I have developed for those among them who like to try it—it is not a required department course—has been listed in our College catalog as Physics 120, "Classics of Physics," and described briefly as follows: "Selected case studies in the rise of modern physics, from mechanics in the 17th century to the origins of relativity, nuclear physics, and quantum theory. Lectures and guided reading of original papers and of secondary literature, some seminar work, and a term paper."

The prerequisite is an introductory course in college physics. But most of my students have more than that, being juniors and seniors in the physics department or, in some cases, students in the history of science program whose special interest is in the physical sciences. It is frankly not a soup-to-nuts course of the rise of modern physics; there are other such courses in town which they can take if they wish. Rather, I deal with cases which illustrate main turning points in the history of the physical sciences; they are classics which hang together, although there are gaps in time between some of them since I try to give as much time as possible to the twentieth century. The course has been changing from year to year because I like to include cases on which I am working or about which I am gathering ideas for a future paper.

To give an idea of the kinds of cases dealt with in this selection from the classics of physics, let me list the topics selected in a recent year. To begin with, Galileo's mechanics in the *Two New Sciences*. The Newtonian contributions, as shown in parts of the *Principia* and the *Opticks*. The rise of electromagnetism in Oersted, Ampére, Faraday, and, to a small degree, Maxwell, with special attention to the rise of *Naturphilosophie* in the case of Oersted and Ampére. The wave theory of light, focusing on the publications of Thomas Young. The fundamental papers on the discoveries of X-rays and radioactivity, with attention also to the role of these radical innovations on writers outside science, ranging from Henry Adams to H. G. Wells. Then the basic papers of Planck on the introduction of the concept of the quantum of energy, Ernest Rutherford on the nuclear atom, and Niels Bohr on the quantized atom. Two segments on Einstein's special relativity, followed by the discovery by the Fermi group of artificial radioactivity by slow neutrons, and finally the reformulation of quantum physics by Heisenberg.

An important part of the course is the work the student does on a major term paper, for which he or she is to select a classic case outside those we cover in class. For many students this is a big challenge, since so few of them have ever been forced to write a major paper; but it is also a high point, both for them and for me, since I meet separately with each student repeatedly, first to help select the topic, then to help prepare a bibliography, to go over a draft outline, and if they wish to go over the draft paper itself. Over the years, many of these papers have in fact resulted ultimately in publications by these students, since I urge them to investigate a topic that allows originality.

Topics of some of the student papers of past years on twentieth-century science included these: Hermann Minkowski's version of special relativity; the scientific career of Louis de Broglie; the experimental checks on the equivalence of mass and energy; the Pound-Rebka experiment on the equivalence principle; the Curtis-Shapley debate on the existence of galaxies beyond ours; the rise of physics in Italy; G. N. Lewis's theory of valence; the rise of quantum physics in the United States; the origin of cosmic ray research; the reception of general relativity in English-speaking countries. And among papers dealing with earlier periods: "Du Fay's Contribution to the Experimental Method"; "The Great Vector vs. Quaternion Debate on Maxwell's Electromagnetic Theory."

It should be evident that one of my main aims in this course is to allow students for once to see the field of physics in the manner which is prevented by the more usual pedagogy that has forced us to cut up the field course by course. The usual deconstructive approach makes it easy to teach, but it is completely at variance with the more organic use of physics in any real research situation. It is also at odds with the multidimensionality of understanding any development in the history of science. By contrast, a course of the sort I am describing can hope to provide a more coherent view of the field than the various separate fragments on, say, mechanics or optics or electricity and magnetism or atomic physics, and so on. Dealing with actual, major papers allows one to analyze not only the content of a real research paper, in which not one but several fields of physics are likely to be present, but also the historical situation in which the advance or the failure occurred, and the various forces inside and outside science that helped to shape the work and its initial acceptance or rejection. Where possible, we can bring in some aspects of what Einstein called "the personal struggle," and also the degree to which work in twentieth-century physics is methodologically similar or different with respect to that of earlier periods.

As texts for this course I use each of the classic papers, sometimes with secondary publications analyzing these, as well as a number of required and recommended books, including the two-volume series of Emilio Segrè, volume 1, *From Falling Bodies to Radiowaves*, and volume 2, *From X-rays to Quarks*. I recommend four other books, although they are not required: Einstein's *Ideas and Opinions*, Peter Galison's *How Experiments End*, A. R. Hall's *The Revolution in Science, 1500–1750*, and Daniel Kevles' *The Physicists*. Additional reading assignments in a great variety of books and articles are made available to the students on a special reserve shelf.

With all these good resources, there is a lot of reading to be done, and this is not always easy in today's student culture. What provides an extra amount of energy, though, is that, usually for the first time, the student gets in touch with the real thing, the original physics research paper. Exposure to the classics emanates a sense of authenticity and excitement, and of course it also sometimes gives models for scientific thinking and presentation.

I should remark that I have been aware that such a course could be put into place perhaps a little more easily at some colleges than at others. In our case, the hospitality of the physics department surely owes something to the historic sequence of professors who have been interested in the history and philosophy of science—persons such as P. W. Bridgman, Philipp Frank, Einstein's successor and biographer, Edwin C. Kemble, who brought quantum physics to the United States while a graduate student himself, and Sheldon Glashow and Steven Weinberg, each of whom has been writing on the history of science. From beyond our own department we see of course the historic shadows of George Sarton and of James Bryant Conant, who taught a history of physical science course himself while a president of the University.

The history of science can also be put to the service of science teaching in other, more usual ways, ranging from reading courses for physics students to guest lectures on specific historic topics, to a Freshman Seminar with titles such as "The Life and Science of Einstein." Having endeavored to keep the main examples from my experience as operational and instrumental as possible, I confess the hope that these experiments will give courage to others to do similar or better things. Yet I must conclude with a speculation. Our educational methods are under new scrutiny, and the sciences themselves are under various sorts of attack, not least for their presumed isolation from other human concerns. It may therefore well be that at this point in history it is more appropriate than ever to infuse our teaching with the sense of historic connection between each particular science and the other aspects of our culture.

References

1.   Stephen G. Brush, *The History of Modern Science: A Guide to the Second Scientific Revolution, 1800–1950* (Iowa State University Press, 1988); Edwin C. Kemble, *Physical Science: Its Structure and Development* (MIT Press, 1966); Gerald Holton, *Introduction to Concepts and Theories in Physical Science* (Princeton University Press, 1985; revision prepared by Stephen G. Brush).
2.   The kind of titles of papers students have typically written during one academic year in response to this assignment are given in a more extensive description of the course in Chapter 15, *Thematic Origins of Scientific Thought* (1973 edition, op. cit.).

Faraday's
"Advice to a Lecturer," Updated

Every day, millions of students in schools and colleges the whole world over sit in science classes, listen to their instructors and watch them perform time-consuming demonstrations and lecture-table experiments. What good do these presentations do? What is the mechanism by which they teach? What *else* is being conveyed at the same time? What are the different styles of presentation? Is an actual human being needed as lecturer, or can his or her function be fulfilled equally well or better by some surrogate such as a filmed image?

From the very outset I am aware how little is known about this subject in any fundamental way, and how little has changed since Michael Faraday wrote "Advice to a Lecturer."[1] This lack is rather astonishing, because for those who have been working on the psychology of the learning process the demonstration in a classroom should serve merely as a special case—in principle, even a particularly controllable and testable one—of a more general problem, the problem of how knowledge, technique, and attitude on any subject are conveyed by one human being when another person or a group watches acts of demonstrance. This is, after all, one of the most frequent methods of communication, whether between teacher and student, parent and child, or actor and audience.

Without attempting a rigorous definition, we can say that in the general case, a demonstration always involves one person who knows or feels what some others do not yet know or feel. This person wishes to communicate this knowledge or these feelings to others vocally, by gesture, or through symbols or external objects or devices. The conception of demonstration as an instance of communication in general makes our discussion of the special case of scientific demonstrations more widely useful.

To be sure, the demonstration-centered science lecture course does not have a long history. Writing of the practice obtaining in 1847 when the Cambridge (Massachusetts) English high school opened, its first headmaster, Elbridge Smith, noted: "In science, the instruction was wholly by catechism. Here were illustrative diagrams in the textbook which might, or might not, be explained and might, or might not, be transferred to the blackboard....Not a single particle of apparatus or a book of reference, except the Bible and possibly a dictionary."[2] The wealthier colleges and schools had started to show demonstration equipment to science classes earlier; but by and large Smith's description also applies to college physics courses in the United States at that time. The demonstration apparatus as a teaching tool was still considered important mainly for grand occasions (one thinks here of the magnificent orreries and tellurians), or for popular lectures, addressed to a wide audience, and largely built around surprising or impressive demonstrations.

Since those days, physics instructors have by and large not achieved renown for excellence of presentation. Indeed, many of the best scientists have been bad lecturers—some of them embarrassingly so—as was also earlier the case with Kepler, Young, Helmholtz, and Rutherford.[3] There seems to be no reason why a good scientist should be a good lecturer, not to speak of being a good lecture demonstrator. Conversely, some of the best scientists have reported that during their student years their physics lectures did not attract them, as, for example, Einstein reported in his "Autobiographical Notes."[4] Indeed, one suspects that lecture demonstrations may well offend the most perceptive students. They are likely to see through the fundamental misrepresentation in the typical lecture-table "experiment," in which a subtle and beautiful phenomenon is distorted almost beyond all recognition in order that the ephemeral visual clues can be amplified for the benefit of people seated at a distance. They must be aware of the fact that demonstrations tend to wrench phenomena from their natural context in order to make a "main point" stand out clearly before the average student. And the most interested and intuitive students may be very uncomfortable when—to cite an example that is so typical— attention is displaced from the real effect to a substitute or analogue, as when an instructor uses a model (e.g., a mechanically agitated tray of steel balls) as the means of discussing a basic phenomenon (e.g., Brownian motion) *without* giving the student a look at the actual case itself.

These three faults of scientific demonstrations—distortion, dissociation, and displacement—will be discussed in some detail below. They are indicators of the gap between physical reality as the working physicist encounters it and what the instructor in the large lecture room often is content to exhibit. It is precisely this gap which the demonstration, among other teaching tools, is usually supposed to be helping to fill; my feeling is that, on the contrary, too many demonstrations tend to increase it. Typical of many reports in the literature to this effect are the findings of an interesting experiment in which a number of instructors from the humanities were asked to attend lectures in the sciences to provide feedback to the lecturer.[5] They had by and large a confusing experience; for example, Sheila Tobias reports:

> At the heart of the lecture on waves in elastic media were [Professor Isaac] Abella's carefully prepared demonstrations. But some of the participants had real problems with these. One summarized his reactions in this fashion:
>
> "There were two types of demonstrations for us: what we choose to call the clarifying demonstration, such as the passing of a wave along a slinky; and a confirming demonstration, one that requires us to see something that is either moving too fast for us to follow or that requires a level of understanding we do not yet have.
>
> "One question the demonstrations did not answer for me was the relation of the formulas to experiment. Sometimes it seemed that the initial and 'crucial' experiments led to the discovery of theory; at other times one felt that the experiment merely illustrated a known formula which is really quite different: the old positivist distinction between the 'logic of discovery' and that of 'justification.' In any case, the question I kept jotting down was: how are the formulas derived?
>
> "Do the experiments *reveal* them? Do they *confirm* them? Or do they *illustrate* them to subsequent generations? Or all three?" (pp. 40–41).

To symbolize this problem, we may mentally juxtapose how a working research laboratory looks—not prettied up for the photographer, not made into a "demonstration," as ever so many published pictures of research laboratories are, but one that leaves unretouched the seeming chaos that strikes one on entering most laboratory rooms—and, next to it, a picture of a class in an introductory college science course. There, the students are facing a lecture table on which there is some apparatus; but above all they are watching the lecturer-demonstrator, demonstrating himself.

The successful demonstration experiment would help us *prevent* the student's eye coming to rest merely in the plane of the lecturer; instead, the student should be able to look through the aperture that the lecturer creates into the world of the laboratory or into nature itself. To the extent that the demonstration and the lecture fail to do this, they are not fulfilling their avowed functions. No one has claimed that the lecture can ever be a direct shortcut to the thinking of the trained scientist at the desk or in the laboratory; but a lecture is justified only if it does increase the degree to which students know and feel science as scientists themselves understand the subject.

I have been stressing here the most ambitious task of demonstration lectures, and one which is getting more difficult rather than easier with passing time. The difficulty of using demonstrations to good effect is increasing for two reasons. One lies in the nature of the advance of physical science. We have long passed the point when by means of pulleys, magnets, and electroscopes one could demonstrate the current state of science—when, occasionally, a fundamental discovery could be made in the course of preparing or giving an elementary lecture, as was the case with Heinrich Hertz.[6] As the physical events of the research laboratory have gone further and further from the "naked-eye" stage of science, so has, necessarily, the physical reality in the lecture room where physics is being taught. In the demonstration of, say, a solid-state or relativistic effect, one requires not only far more sophisticated *equipment* than had been used in the past, but far more sophisticated *eyes* than many students are likely to have, and sometimes also more sophisticated hands than those of many a lecturer or demonstration curator. It is true that one must start with those tried and true demonstrations which lead properly from the simple concepts to the more difficult, and so one must demonstrate Newton's laws, ripple-tank interference, and the like. But is it not also true that difficulties with demonstrations of more up-to-date physics are forcing many lecturers to stick to their old demonstrations—not because they are still needed, but because they can be done more easily?

A second reason for the increasing difficulty with demonstrations lies in the practical realm. New physics lecturers are, more and more often, men and women who have passed through no substantial period of teaching apprenticeship. They may face a large class for the first time when they are put fully in charge of a course. Nor can they count on skillful and reliable curators to help behind the scene. Except for cases of rare luck, we have long ceased to have worthy descendants of a Robert Hooke, a James Watt, or a Michael Faraday, all of whom were first-rate preparators of demonstrations. Moreover, the lecture-equipment industry in the United States is not unduly disturbing itself with new ideas and aids either.

As a result of these trends, we have been moving into the era of the "Automation of Teaching" and the "Technological Classroom," in which the role of the live demonstrator-lecturer will be considerably attenuated. Some scientists are on record that they see no reason why the lecture portion of large elementary courses in schools or colleges could not be handled through TV or film.

To be sure, certain films or well-rehearsed and well-equipped science lectures are very useful. There is no question that all bad lecture demonstrations (and lecturers) should be replaced, if necessary, by good audiovisual surrogates, just as we might let much of the necessary rote teaching be done by means of computers and other programmed devices. And even in the classes of the good demonstration lecturer, the usefulness of film, closed-circuit TV, etc., is larger than many have thought in the recent past.

But we must not get caught up in our usual optimism that any new, promising development should be extended as quickly as possible from horizon to horizon. We have several reasons for thinking carefully before we settle for the "Technological Classroom." The human demonstrator is obsolete only if we compare the few good films or TV programs with a mediocre or bad lecturer. Second, the average lecturers may not be beyond substantial improvement if we find a sound theoretical basis for understanding their function—an argument in favor of a search for some new ideas and speculations of the kind presented here. And above all, a number of important values are conveyed, in the interaction between the live lecturer and the student in the same room, particularly through the scientific demonstration, and these values depend on communication being initiated by an actual person rather than by a surrogate. These values, and the methods of communication, need to be recognized and preserved. To do this, we must go beyond what one might call a classical theory of the function of the demonstration lecturer.

The Classical Theory of the Lecturer

The classical model of lecturer-demonstrator in a science course is well established, though there is a variety of types. The prototype is perhaps described best in that charming booklet gathered from the early writings of the master demonstrator, Michael Faraday, *Advice to a Lecturer*. His counsel is what one would expect from a nineteenth-century expert. The apparatus should be carefully laid out, and the demonstration should have a certain magnitude, appropriate to the point to be made; for when the lecturer "calls your attention in a particular way to a decisive experiment that has entered his mind, clear and important in its application to the subject, and then lets fall a card, I turn with disgust from the lecturer and his experiments" (p. 6). His actions should be "slow, easy and natural." His discussion should be lucid and to the point: in Faraday's study at the Royal Institution, I am told, the two main objects that greet the visiting lecturer preparing for his task are a big, beautifully formed quartz crystal and a large container of barnacles; they remind the lecturer to be crystal clear and to stick tenaciously to the subject.

The lecturer, Faraday reminds his readers, should behave like a "gentleman," and the members of the audience should be treated as if they were, too. Faraday repeatedly comes back to the need to "evince a respect for his audience. No accident…should disturb his serenity or cause variation in his behavior" (p. 1). "Apologies are to be made as seldom as possible and generally only when the inconvenience extends to the company" (p. 5). No lecturer should descend "so low as to angle for claps and ask for commendation" (p. 5). And above all, "a lecturer should exert his utmost efforts to gain completely the mind and attention of his audience, and irresistibly make them join in his ideas to the end of the subject" (p. 3).

In short, the classical demonstration lecture is a unidirectional process, executed with an almost Victorian sense of purpose and decorum. It is orderly, tasteful, and well programmed—and in all these respects, it is replaceable (but for the magnetism of Faraday himself) by a well-made film. For the key aim here is the "effort to gain completely the mind and attention of his audience." This tight grasp is indeed exactly what film or TV can achieve better than any other medium.

In this century, the classical theory has been changed by the recognition that the lecture demonstration is only one in a system of connected course elements, one part of a whole spectrum of teaching aids. Video cartridges, which the student can use in the library or take to the dormitory, are one example of the extension of demonstration lectures. The student's handling of the demonstration equipment itself after a lecture or in a conference or a laboratory section is another. Corridor demonstrations in which lecture-table apparatus is made available later to the student is yet another. Kits by which the student can repeat or go beyond experiments taken up in lecture or laboratory have been introduced also. These and others are healthy developments by which the course becomes less of a hypnotic, one-way, lecture-dominated process. The student can be drawn into the process and become more than a passive consumer.

These changes in the age-old model were partly prompted by the realization that what the student retains from seeing and hearing alone is so poor that the lecturer's spoken word should be accompanied by other stimuli, such as demonstrations, and that learning is enhanced if the students participate and manipulate things themselves. Moreover, the classical theory has been influenced in the same direction by the increasing stress on conceptions such as "creative thinking" and "teamwork" in the classroom—conceptions which reflect a general sociological change more than a mere pedagogic trend.

Natural Complexity vs. Demonstrable Simplicity

A thorough *non*classical theory goes beyond these modifications and starts from a recognition of what a lecturer cannot do well, and what a technical substitute can do even less well. If the main purpose of the demonstration is to help us bring physical, professional reality into the classroom, to show what science deals with, how its work is done, by whom, and why, one must first explore the extent to which this aim is inherently not fulfillable.

One obstacle lies clearly in the fact that the phenomena of current scientific interest are very difficult to transport into the classroom; that, for example, in a laboratory investigation, the sense stimuli that really count are rare or quiet compared with the total sum of stimuli clamoring loudly for attention.

But perhaps the deeper difficulty is that one knows any one thing X in science (or for that matter in any other field) in a professional way only if one knows also the connections of X with many other things Y, Z....The facts and experience of science are not atomistically given, but reside in a learned pattern of connections in the subject field. For example, the truth and beauty of the fact that the electron has a constant charge is at most only hinted at by what a student can see by watching a re-enactment of the Millikan oil-drop experiment. To understand the meaning of the observed motions of the drop requires two profound steps: first, connecting the motion with such other ideas as the measurement of the density of oil and of air, Stokes' law and its range of applicability, the ionization of gases by radiation, the construction of regulated voltage supplies, the statistical treatment of data, and so forth; second, the constancy of the fundamental electronic charge discovered in the oil-drop experiment has to be fitted into a network of ideas involving quite different experiments, including the constancy of mass transfer in electrolytic dissociation, the constancy of the radius of curvature of cathode rays in magnetic fields (as well as the departure from constancy of the e/m value at high beam speeds), the reappearance of the same value of elementary charge in photoelectric phenomena, charge conservation in pair production, etc., etc. And each of these ideas itself is similarly at the center of its own net of other conceptions. We now realize that a demonstration at best shows only one knot, rather than the threads which point to and account for the central points in the whole network of ideas.

Here an analogy can short-cut a discussion that belongs in the philosophy of science. Anyone who has seen the set of tables in the Ishihara Color-Blindness Test remembers that each table consists of a circular field, about four inches in diameter, filled with separate dots of different colors. In one table, for example, the normal eye may see a digit 9 formed by dots of one color in the field of other dots. But the color-blind person will see a 3 or a 7 or no number at all, depending on the form of color-blindness. What the normal as well as the abnormal eye sees is, of course, a learned pattern, just as any "fact" or portion of subject matter is of importance to the physicist because it is perceived by him to fit into a larger, learned pattern. It is not the fact, the "isolate" itself, that makes it interesting to him, any more than a single dot on the Ishihara table is of significance. What makes an isolate meaningful is, in the first place, that it is perceived to lie, as it were, along the line that has the form recognized as a 9 in the visual field of the table as a whole; and, second, that from a different point of view, the *same* isolate can also be regarded as lying along a 3 or a 7.

But precisely here is the difficulty for the student: the number of connections that characterize most isolates of physics are immense; this is, after all, what is meant by saying that physics is a highly structured area of knowledge. In order to digest and benefit from the instruction which is offered, the student therefore tends to do what any digestive system does, namely, first break up the complex structure into less complex parts. Instead of dealing with the digit 9 as a whole, the student concentrates first on some part of the curlicue. And the lecturer, too, helps him to pick out what is "seen" easily, as it were, against a blank background instead of the initially confusing background of other colored dots.

This is where the lecture demonstration enters. The motion of an electric charge in a field cannot, to the uninitiated, be explained or demonstrated by actual examples where this conception counts—either in "real" nature (e.g., motion of protons in the field around the sun) or in the laboratory (e.g., motion of protons in an accelerator). Instead, the lecturer might show the e/m experiment of bending a fine beam of electrons in a large, evacuated globe between circular Helmholtz coils. It is a beautiful "experiment"—for the few students in the front rows who can see something of it—but, like most lecture demonstrations, it is of necessity and almost by definition a carefully adjusted, abstracted, simplified, homogenized, "dry-cleaned" case, far removed from the physical situation in which charged particles traverse fields, either in out-of-doors nature or in the contemporary research laboratory. As presented on the demonstration table, it is as if all the blue, red, yellow, and purple dots in the Ishihara color plate had been removed so as to let a continuous line of green dots stand out against a blank background, and then much of the digit 9 itself had also been cut off to focus attention on only part of it. The trouble is, of course, that this portion of the original pattern by itself is far removed from the interesting information; the original, complex figure 9 that can also be a 3 or a 7 has disappeared, and in its place there stands something that might be mistaken for a naught.

Experimental Reality and Didactic Reality

I have noted that demonstrations are necessarily simplifications of a natural complexity; they are almost always distortions of the physical situations with which the physicist works. We may thus, for the purposes of this essay, distinguish between two kinds of reality. One can be called *experimental* reality, namely, that of phenomena and events in their full context, embedded in rich complexity, burdened with secondary and tertiary "side" effects that are not suppressed and forgotten—the 9 against the full-spectrum color background that also implies the 3 and the 7; the electron that is not merely a thermionic emission hurled into a homogeneous magnetic field but that also may be a net charge on an oil drop, or a photo-electron, or a decay product in a bubble chamber; the force on a moving object due not only to gravity but to the resultant of gravitational and electrostatic forces as well as those of air currents.

For the purposes of evaluating the usefulness of pedagogic techniques, I shall introduce here the heuristic notion of "levels of reality," and identify the experimental reality of the research laboratory as a "first level" or a "first order" of reality. Below it, at the second level, there lies the arranged, *didactic* reality of the lecture room. At its best, it can make a wonderful show and teach a great deal—as, for example, microwave optics equipment. But there are not many examples. After a steady diet of didactic demonstrations, a musty, stale atmosphere settles over the lecture, as on almost everything else done in the course: on the student-proof laboratory setup, on the aged exhibit, on the obedient turn performed by a captive and tamed phenomenon—the ancient finger exercise instead of actual music.

What I am stressing here is not an argument against good demonstrations on the level of didactic reality, but rather an argument for supplementing these by introducing first-order reality far more frequently than is generally done. *Even at the risk* of bringing into the classroom, occasionally, a sample of reality on too high and therefore too complex a level for the students' full comprehension, I would urge three remedies for a class surfeited with didactic reality. One is that students should see (and, through the laboratory, participate in) experiments that are not precleaned for them, where the surfaces are not smoothed and planed, the connections not supplied in just the right number, and stray fields not shielded against.

Second, in this light one sees the increased importance of visits to on-going research laboratories. Every large experimental installation on or near the college campus should be obliged to have a well-designed area in which students can come to visit and watch, even if it has to be at a distance and from a glass enclosure.

But as important, even essential, as such visiting areas and visitors' days are, I have my eye also on the goal of taking the class into not one, but many functioning laboratory rooms. Thus, following the lecture discussion of the motion of charges in fields, the conference or recitation section might be held in an instructor's molecular beam laboratory, even if it is crowded and the apparatus can be shown only briefly to illustrate some of the course discussion. We must experiment more with the occasional transfer of the class from the antiseptic section room into those messy places where what one sees was not made merely to be shown to students, where experimental reality teems and squirms in its florid and maddening abundance.

Third, we should bring into the regular classroom, as often as possible, equipment and pieces that function in commercial technical apparatus or, better still, that fulfilled some purpose in actual research. We may perhaps be allowed to call them *relics*. We do not mean to turn our backs on well-designed, didactic exhibits, yet occasionally there must enter into the science classroom a piece of equipment that carries with it an atmosphere of authenticity, a sense of what lies at the end of all this pedagogic experience.

Therefore, instead of tying equipment up irrevocably for research only, and particularly instead of throwing it out after the research is completed, we should think whether it may not offer a valuable opportunity to tie a point being discussed in a science course to some research actually done at that institution by that very physics department, by one of our own colleagues or ourselves. Even a single component—say, a piezoelectric crystal or a grating—is better shown to the learners *in situ* than torn from its chassis or preserved in its wrapping. Moreover, the authentic equipment can convey useful information over and above the scientific one; for example, it reminds the student in a meaningful way of what a college is all about and what else his instructors are doing.

Hence, perhaps a dozen times a year, relics that have been involved in respectable research should be introduced into the discussion in the classroom, if only briefly. The collection of local relics need not put a strain on the facilities, but it does mean that someone in the department should keep after his colleagues to supply (and label) such equipment and should speak up for enough storage space in the lecture-preparation area.[7] It is worth it. I can attest to the fact that the introduction of relics can have an electrifying effect upon the class. Thus, our classes are always intrigued by the simplicity of the shop-made equipment with which E. M. Purcell, for some early research, obtained a narrow beam of beta rays; we now use the source in lectures to demonstrate deflection by means of magnetic fields. Or, at another time, we pass around the class a varnished cube of uranium oxide which E. C. Kemble brought back from the "Alsos" mission to Germany toward the end of World War II in search of the reactor built by the Germans in their ill-fated nuclear weapons project. The students can at last touch something; can feel with surprise how heavy the cube is; can calm their worry by recalling that a long half-life corresponds to a relatively low disintegration rate; and so forth.

If such nonpedagogic equipment was involved in some research important in the progress of physics, these pieces will carry with them (through the way we present the matter) a reputation similar to the aura that surrounds significant works of art. Even a well-chosen piece that was part of a more routine investigation has its uses. For instance, it can convey the same respect that any example of good craftsmanship evokes. And at any rate, these exhibits of nonpedagogic items are necessary tokens of first-order reality—necessary for many reasons, but none more important than the easily neglected fact that the student, unlike the instructor, usually cannot even know that there *are* different levels of experience. Hence the instructor can explain the difference by means of examples.

Depiction of Lower Levels of Reality

In the last section I have conspicuously left out what is perhaps the easiest way of representing physical reality in the classroom—namely, by videotaping a sequence in nature or in an actual laboratory, or by the use of slides, photographs, etc. All these are, of course, immensely important parts of the spectrum of lecture aids, and they have become as natural and ubiquitous as are records, tapes, and sound-reproduction equipment in a well-run music course. Almost everyone seems to use the films of the collapse of the Tacoma Narrows bridge and of events recorded by NASA crews during space travel. I should confess that my own favorites include the brief films, originally made by the Project Physics Course as three-minute film loops and now on videotape, of phenomena in nature (such as crossing a river against its current).

But it is important to realize that with films, TV, slides, and photographs, we have gone down to a third level of reality. We are now speaking of watching *depictions* rather than the events themselves. Furthermore, these devices are used in almost every case for depicting on the two-dimensional screen not even what has happened in first-order reality, in the ordinary physical situation of the laboratory or in nature, but what has happened in a staged, didactic performance. Usually, we are dealing here with, say, movies of phenomena in a ripple tank—a device invented by Thomas Young for his lectures to laymen at the Royal Institution, and since that day virtually never used for anything but pedagogic purposes. Such films are, therefore, representations of representations. Recognizing on which level these media operate will warn us that such teaching aids, if used too extensively, can take us away from the avowed aim of the lecture demonstration, which was to help close the gap between the classroom, on the one hand, and physical reality experienced and understood in its full context, on the other.

At another order below depiction, we come to a level of classroom reality on which we find *models*. Now the avowed contact with first-order physical reality has been almost completely broken, and the device that is shown functions analogically. The student is looking at a translation, crafted in wood and metal and paint, at an analogon, rather than at the phenomenon itself or at a depiction of it. These kinds of "fourth-level" devices share the characteristic that the physical forces and laws by which they actually work are not the same as those that are being "demonstrated." A sharp discontinuity in meaning and authenticity, therefore, exists between this and the higher levels of reality. Watching a movie of Brownian motion in a colloidal suspension is, in terms of perceived dramatic impact, less impressive than seeing Brownian motion itself through the microscope's eyepiece; but a model for this phenomenon, using agitated metal spheres, is less striking than either of them. It still can be very useful, if one is clear on what level it operates.

A special case of model-based presentation is computer-generated animation. This has become a very seductive tool; but its most useful aspect is the presentation of physical events that are inaccessible by other means and that therefore share in the property of "*Gedanken*"-experiments; examples are the appearance of objects at relativistic speeds and the delightful imaginary trip, in the film "Powers of 10," into outer and inner space.

Moving on to the next-lower, fifth order of reality, we come to the kind of teaching aids that form the bulk of most lectures: graphs and slides, writing and talk. They may be termed *condensed coding*. A slide made by time exposure of a particle in Brownian motion is a shorthand device that eliminates the time dimension by representing the whole sequence of happenings at one glance. This is, of course, even more true of a sketch of a zig-zag line drawn between points corresponding to a sequence of observed positions of a particle in Brownian motion.

These projections, graphs, written equations or phrases, and, indeed, the lecture talk itself, all convey information by indirect means (rather than by exemplification); they are a form of condensed coding concerning possible happenings, similar to the projection of a musical score instead of the performance of the piece itself. The analogy I wish to stress here is that to the skilled musician or musicologist, scanning the score is closely parallel to the experience of an actual musical performance. In a sense the same applies when a physicist reads someone else's skillfully presented paper on results in his own special field of research in a professional journal.

But this argument cannot be extended to the student. The lecturer might easily say, "This is how waves interfere," as he projects two superposed transparencies depicting sets of concentric black semicircles; but the student, who is unable to make the shortcut with the same intuitive facility from coding to a phenomenon, sees first of all merely a multiplicity of more or less well-focused lines on the screen. Many students make the transition from the abstraction to the phenomenon only slowly and with difficulty.

There are more remote degrees of abstraction below those cited so far. At the sixth level is thinking about scientific problems without ordinary language, in the language of direct and therefore metasymbolic abstraction of which Einstein spoke in his reply to Jacques Hadamard.[8] At this level, on which thinking in all scientific fields of any originality proceeds probably much of the time, the words, graphs, and equations have disappeared. And below that, at the seventh level, is direct apperception of a complex situation in its entirety, including the case of suddenly having a solution rise into one's consciousness when one is not even aware of having made the effort to think of the problem.

Table 1 lists the orders of "reality" from experimental and didactic down to abstract and unconscious. A well-trained physicist, perhaps the lecturer of the course that we are discussing, has long ago learned to "see through," or tunnel rapidly, from the first order of reality to lower levels. It is, after all, on the lowest levels that he "understands" his subject matter and "knows" the many implicit connections of the isolate which he is discussing, rather than on the higher levels that take so much time in the presentation of the lecture demonstration. Understanding scientific materials, and perhaps understanding anything, depends on a cybernetic act by which one leaves the first level, the realm of the perceived phenomenon

TABLE 1. Orders of Reality in Scientific Presentations

1. Experimental reality
2. Didactic reality
3. Depiction of phenomena (film, TV)
4. Analogon (model, animation)
5. Condensed coding (graphs, equations, words)
6. Metasymbolic abstraction
7. Unconscious or intuitive apperception

that poses the problem initially, and transfers the internal dialog to the lowest levels; the decisions reached there allow one then to go back to the field of phenomena itself—to look, so to speak, for the heretofore unrecognized 7 that stands behind the initially obvious 9. It is in this manner that we discover and learn, that we decide what it is in the chaos of events which ought to be looked at and called first-order reality, that we select what is worthy of our study and of communication to our students.

It follows from this hierarchy, perhaps surprisingly, that the use of demonstrations pitched at the intermediate levels may not always be an asset. For some students, the appreciation and skill of the quick "tunneling" process from the uppermost to the lowest levels may be not advanced but retarded by stressing intermediate presentations. One is reminded of Lagrange's decision to keep diagrams out of *Mécanique analytique*, as if he feared they would make the argument too concrete and specific. An opinion has been advanced by a number of thoughtful people, such as Edwin H. Land, that at least for certain students, creative learning may not occur in a classroom situation nearly so well as when the student is from the beginning put into the laboratory and is allowed to work on problems whose natural complexity is not hidden by a didactic, prior separation of the elements.

Further Pedagogic Consequences of Recognizing Orders of Reality

A number of further consequences follow from the recognition that in the physics classroom, the "reality" of the subject matter can be presented on any one of several levels.

(a) It seems to me plausible that the closer the demonstration is to the first level—the experimental level of reality, which is, after all, the subject

TABLE 2. Presentation of Interference on Different Levels of Reality

Level	Example of Demonstration
1	Exploration of interference pattern set up by industrial or research device
2	Operation of ripple tank or other pedagogic device
3	Film or TV image of (1) or (2) above
4	Simulation of interference phenomena by an analog device
5	Drawing, projected or on a blackboard

of actual study by the working physicist, the source of his understanding and his puzzles—the greater is the pedagogic effectiveness of the demonstration, as long as technical obstacles of size, visibility, etc., do not stand in the way.

To translate this postulate into practice: during the introduction of the concepts of interference, for example, one should rely not on diagrams or models or films but on exhibitions of actual, first-order cases: exploring an interference pattern by means of a suitable probe sweeping in front of a mounted loudspeaker or around a transmitting antenna; asking the class to look at the light from a bare long-filament lamp through the space between the forefinger and middle finger held close to the eye to see the interference fringes; showing interference patterns from thin films, etc.

Later in the same set of lectures there will also be opportunity to discuss interference at each of the other levels: on the didactic-reality level, one may be using a ripple tank, or the equipment for interference of centimeter waves radiated from one or two dipoles; and so forth. On the third level (depiction), films may be used, both of natural and of pedagogic events that illustrate interference. On the fourth level (analogon and model), interference can be shown by means of machines in which waves are represented by bent wires, etc. On the fifth level, diagrams, graphs, etc., are available on screen or blackboard. This whole spectrum is represented in Table 2. But the theorem stated at the outset would ask us to stress, emphasize, favor, lavish particular care on, begin with, and repeatedly return to examples at the highest levels.

In the absence of data, one can only speculate on the question whether for pedagogic purposes a natural sequence exists from the first level down. The natural sequence appears to me to be generally from the top level to the lower ones. This is probably contrary to the order obtaining in most lectures today (i.e., one tends to start with a quickly drawn sketch on the blackboard or a simple model). Yet the vertical progression from observable raw phenomena to abstraction and explication is usually prefer-able in the introductory science class, according to the theory developed here—above all because of the desirability of anchoring the concepts be-ing developed as closely to their complex natural habitat as possible. If this order is adopted, one may well return at the end of the sequence to re-peat in greater detail the first, conceptually richest type of higher-order demonstration.

(b) Lecturers should make clear to themselves and the class on what level a demonstration is presented. Students are generally not so agile that they can distinguish quickly and clearly between the analogon and the re-ality it tries to represent. The difference can be spelled out briefly enough, but it should be explicitly noted.

(c) If we take the primacy of upper-level presentations seriously in planning demonstrations, we should recognize that the same need exists also with respect to the problems that are assigned to a class to be worked at home. All too often these problems are simplified versions of physi-cally realizable situations (e.g., the immovable, frictionless plane). These have, of course, the value of simplicity for student as well as for grader. But at least occasionally, experiential reality in its greater complexity and richer context should enter even these problems, and instead of giving di-gested, reduced data of an idealized *"Gedanken"*-experiment, we should give the raw record of an actual (even if didactic) run. For example, give not just the data for distance and time but at least the stroboscopic photo-graph depicting the motion; and better still, take the photograph as part of a lecture demonstration in the first place; and even better, let students take it themselves as part of their laboratory activity.

(d) Depictions and analogies are translations. Therefore, one must be watchful for incomplete rendition, mistranslation, and unintended mean-ings, as in all other such cases. All translated messages tend to suffer from nonlinearity, limitation of band width, and noise. In terms of our discus-sion, for example, a mechanical analogon of wave motion will leave most students with the implicit feeling that there is no difference between small- and large-amplitude phenomena; this error is less likely if they can watch ripples and waves directly.

A model, a film, a slide, a projection, a blackboard drawing invariably changes the setting of the phenomenon. An obvious example of this effect is the loss of a natural size scale. In most such cases, the student does not know whether he is looking through a microscope or a telescope. I recall, for example, the talk announcing the successful production of quartz crys-tals grown in autoclaves by a commercial company. A slide of a beautiful crystal was projected, an impressive photograph filling the large screen and spilling over on the wall to each side. Only in the question period did it turn out that the length of the crystal was a few millimeters.

(e) The uses of depiction in particular need to be greatly improved. It is, for example, unfortunate that classes still do not have adequate means for depiction in three dimensions by projection, or by inexpensive individual viewers with adequate optics. We may have to take our cue from the way that the overhead projector became available initially—not through the efforts of the two or three now-ubiquitous educational companies that divide among themselves most of the equipment market, nor, for that matter, through the large specialized optical companies, but through a small independent company that had not previously been prominent in developing instructional devices. Physics instructors will continue to have to do research on educational devices themselves.

(f) When all is said and done, some uses of lecture demonstrations will not fit well into any scheme, but will nevertheless recommend themselves for reasons of their own. Thus, the occasion will arise for some demonstration simply because it is beautiful. The changing interference patterns of light from thin films; or the tracks in a continuous cloud chamber; or even just the back-and-forth flow of energy in coupled pendula—such demonstrations deservedly get more play than may be strictly necessary on pedagogic grounds alone. There must be time enough to gaze at some of the visually fascinating and esthetically satisfying things of science for their own sake.

Another defensible aim is to convey a problem, a paradox, a puzzle. It is particularly appropriate if for this purpose lecture demonstrations are chosen which follow more or less the original experiment in which that problem or puzzle arose (e.g., Oersted's experiment, Hertz's photoelectric effect for polished metal surfaces, C. D. Anderson's photograph of a positron track). Then also a well-placed experiment may help to break the dullness of a classroom routine, refresh the attention of the students, insert some humor. One remembers Leonardo da Vinci warning that one should not judge art when tired; nor should science be so judged. In this category of stimulating occasions belong such demonstrations as the "rocket car," consisting of a CO_2 fire extinguisher mounted on a small, steerable, three-wheeled carriage, which demonstrates the action-reaction principle while the lecturer, seated on the car, is jet-propelled out of the room.

I must confess here my prejudice against such spectacular demonstrations, except on rare occasions. Otherwise, we train the student to expect a show, to have as a matter of habit the reaction with which Diaghilev is said to have greeted Jean Cocteau when they first met. Diaghilev looked at the young man of whom he had heard remarkable things, and said only: "Astonish me."

The Function of Human Presence

In view of the distortion involved in didactic presentations, and even more in analogons, graphs, and similar translations, the question arises why these lower-level demonstrations are in fact pedagogically as useful as they often are. Why is it that we can do so much with models and drawings when the physical reality we are discussing with these aids is kept so far away? The large part of the answer, I believe, lies in the fact that these devices are mediated by an actual human being, the lecturer in front of the student. We therefore must now try to clarify the function of the live lecturer. This really depends, to put it simply, on knowing what the important differences are between human beings and their surrogates (e.g., a projected image of them, their printed or otherwise programmed instruction, etc.).

The success of devices at lower levels of reality is puzzling in the same way as the early success of Shakespeare's profound and subtle plays was puzzling; for the original audience—largely the illiterate laborer in the pit—could not understand the classical allusions or complex historic references, and yet it felt the power and meaning of the play coming across the footlights. To a large extent the visual experience, the movement, light, color, or costumes, helped. But there is something more, and this is added by the actors, the human beings on the stage. The question there, as here, concerns this efficacy of the presence of a human being in conveying a message of great complexity.

We can begin to deal with this difficult and ambitious question if we think first of some of our most successful science demonstrations. Measured by any criterion, two of the most impressive demonstrations I have seen or done are humble. One is an illustration of the law of conservation of energy. The lecturer has previously fastened a long pendulum suspension to the ceiling of the room, the pendulum bob being a large, cannonball-sized iron sphere. The bob is drawn far to one side of the room, the lecturer's head is put immediately adjacent to it. One steadies oneself with one hand against some support and releases the ball. One keeps one's eyes fixed on the audience while explaining that one is willing to risk one's head to demonstrate that the law of conservation of energy applies, and briefly reminds the class how the law is involved in this case. Meanwhile, the ball has been swinging far out and is returning threateningly. (Indeed, if lecturers steal a glance at the pendulum now, they will involuntarily recoil, and so will be made to look foolish.) But—seemingly at the last moment—the ball slows down sufficiently as it regains its original level and momentarily barely touches the head, or is at least again close enough so that it seems to do so. The law of the conservation of energy has given a correct prediction. The lecturer survives.

The second demonstration, which is also well known, concerns the law of conservation of angular momentum. The lecturer, initially motionless, stands on a small platform that can rotate in a horizontal plane and sets a weighted bicycle wheel into rotation with one hand while holding it up with the other in a horizontal plane above head level by means of a rod fastened through the axis. This act causes the lecturer's platform and body to rotate in the direction opposite to that of the wheel. Now one turns the axis of the wheel around 180°—and this strangely also changes the direction of one's own rotation. While inverting the wheel, the demonstrator appears to be struggling with something, may even be in danger of falling off. But though one's dignity may be somewhat impaired by these gyrations, one manages to stay on the spinning platform.

There are two significant points about these demonstrations. First, they are not depictions or models but actual happenings—if not on the first level, at least on the level of didactic reality. They were set up in a classroom especially for the purpose of the show, but they involve pendula and wheels and are tied fairly closely to events that are taking place all the time outside the classroom, as the lecturer makes plain.

But this immediacy is only one aspect of their power. The more important one is the second. The lecturer, by being part of the action right there in front of the class, can show vividly that the pedagogic process and the subject under discussion are *important*. One is willing to risk one's head or falling off the platform (or toppling over in the rocket car mentioned earlier). One certainly is also putting one's dignity in danger. Indeed, there is no more effective weapon for proving a personal interest and involvement to the class.

And this is true not only of these two demonstrations but in a sense of all, with the significant exception of one type, to be noted in a moment. The lecturer's engagement in the manipulation of apparatus right here and now has the quality of taking risks, particularly if it involves the lecturer's own body—risks, it should be noted, that need not come to the student's conscious attention to be perceived. And it is not only my dignity and my physical safety which I am seen to be willing to endanger or suspend. In executing a demonstration myself, I am making at the same time another personal commitment that no surrogate can make. For there is a continuing, implicit possibility that I may somehow cause damage to expensive equipment and in other ways may be led or forced by the apparatus to do outrageous things. This possibility is kept before the class through the unavoidable occasional malfunctioning of an experiment. There is in every spectator of every audience somewhere a savage who by watching the

manipulations or the live stage, the swinging trapeze or the musical instrument, tests whether this is good magic or bad, whether the performer
has mastery over it or not, whether the gods are favorably inclined or not.
Or, to put it in terms of evolutionary advantage, our attention is never
more aroused than when there is a sense of possible danger in the air.
When I meet former students of a course with demonstrations, I regularly ask
them what they remember best from the course. All too often it is an event
when something went wrong. Perhaps this is connected with the finding that
teaching by film is noticeably less effective than by means of a live teacher.

Personal commitment, shown by an implicit willingness to take certain
risks now—this is the element in human participation which is completely
lacking in the operation of surrogates. Little of this element of the unexpected, of risk and commitment, remains in a film; most of it is lost in the
process of "canning." One recalls the skier in the cartoon strip "Peanuts"
who is standing at the top of a small hill, looking down anxiously. At
great length, he says to himself: "I wish this were on film; then I would
know how it comes out." While lecturers who refuse to use good films
from time to time are unnecessarily turning their backs on a major teaching aid, the science class that is conducted chiefly by depiction is, in my
view, in danger of being empty at its very center.

Two Styles

If a theory of the function of the lecturer is to be of use, it should illuminate the variety of styles of lecturing. We can begin to see here the basic
difference between two types: using the terminology Dr. Agatha Petrie
has developed for a rather different purpose, we may discern "Augmenters" and "Reducers." Augmenters are lecturers who enhance or even ex-
aggerate demonstrations in which the unexpected and the threat to body
or dignity are significant elements. They participate vigorously, mount the
lecture table or a ladder, and stress large-scale equipment, smoke, noise,
explosions, etc.; they invite the audience to participate physically and
keep the class quite often in mild uproar; Figure 1 caricatures this type.

The Reducer, on the other hand, is likely to remove himself as much as
possible from the scene as an active participant in the demonstration. He
tends to read his lecture from prepared notes and to use the services of a
curator to help him with the demonstrations, as formerly was the case in
most Continental universities. He stands distant both from the equipment
and from the class; his lecture is carefully planned to minimize the possi-

Figure 1. The Lecturer: Augmenter. Contemporary caricature by James Gillray of experimental lecturer at Royal Institution. Humphry Davy agitates the bellows and Count Rumford stands at the far right.

bility of unforeseen events; his demonstrations are economical and to the point, dry and perhaps even dull. Whereas the Augmenter like to be "hot" and seems to invite risks, the Reducer like to be "cool" and avoids them. If he is worried about the attacks on his personal dignity, he may be protecting himself by being somewhat pompous. While the Augmenter cares so obviously that he makes one sometimes laugh, the Reducer withdraws evidence of his personal involvement so that he makes one sometimes wonder if he cares at all. Figure 2 depicts this age-old type of lecturer.

Lecturer, Audience, and Subject as Coupled System

The engagement of the lecturer owing to physical participation during the demonstration, and the engagement of the audience owing to its only lightly slumbering awareness of the possibility that surprising events may happen—these are by no means the only links by which the lecturer, the

Figure 2. The Lecturer: Reducer. He recites or reads from Galen, and neither participates in the demonstration nor is attended to by his students. (From Johannes Kethan, Fasciculo di Medicina, *1493.)*

audience, and the subject material under discussion are coupled. One such link, probably unique to the science lecture, is that through the manipulation of apparatus or other demonstrations before the audience, an element external to both the lecturer and the audience is introduced: the material equipment itself.

The external object serves to focus the attention both of the lecturer and of the audience upon it. Everyone in the room, even the lecturer, watches the apparatus, contemplates its function, thinks further about the principles it embodies. In a course in the social or humanistic studies or the arts, the main subject matter to which the lecturer and students together explicitly attend is at the bottom always mankind itself, its conditions or relations, potential, habits, or life cycle, tastes or artifacts, etc. But in the science class, the subject to which all persons in the room attend is, as one usually puts it, the function of "external nature." The apparatus on the table is symbolic if not of external nature, at least of the externalization of attention.

From the student's point of view, the lecturer and the material being discussed are necessarily disaggregated. For a conception such as the law of conservation of energy appears in class first as one particularly useful constellation of ideas among the infinitely many possible connections that might be imagined to exist among the isolated elements of thought which the lecturer possesses. Like all such patterns, they were initially learned by being found useful (rewarding) in the experience of the lecturer. In the subject field, therefore, the connections that spell out "law of conservation of energy" are tied to those experiences by which the lecturer has found those connections to be useful, and this is likely to be communicated as part of the presentation to the class.

But in any such projection there is, in addition to the high-fidelity image in the image plane or view field, an area of *low-fidelity* projection surrounding the central image. The student, consciously or not, notices in addition to the law under discussion some of those connections that are projected from the subject field, connections which tie the lecturer's knowledge of this law to the rest of his or her experience and person. In my estimation, it is precisely in this area of low-fidelity, peripheral vision that the student can learn a great deal more than is usually thought. By communicating to the student through this unfocused perception, by encouraging subliminal or "tacit" apprehension along the edges of the view field, the instructor conveys other things in addition to or in conflict with the avowed subject of the lecture. And here I would include the style of mind of the scientist among those things the student absorbs (or rejects). Exposure to the lecturer's thought style in turn helps or hinders the students' understanding of the scientific world view itself, during possibly their only extensive contact with a living scientist.

This low-fidelity area of vision seems to me in particular need of more investigation, for the sake both of more effective teaching of the subject and of fostering a better attitude toward the subject. To give an analogy: it is not uncommon to find students learning French tolerably well in class but then being unable to understand it when they hear it spoken in the streets of a French town. In class, the elements of French language were presented as orderly as a crisp arrow on a clear, two-dimensional view field—whereas the language, as spoken in the streets, is not only a sequence of learned phrases and rules but also a matter of varying speed, intonations, incompletion, and other peripheral intrusions. Similarly, many a student who knows science well by the book is lost when given a problem in a new context. What we perhaps did not let the student see sufficiently, and what confuses at first contact, is the grainy or noisy part of the view field, which in fact attaches to everything we meet in reality.

Two Other Styles

Our model helps us now to discern two other styles of lecturers. On the one hand there are those who tend to project sharply delineated arrows in flat focus on a two-dimensional image plane. One might call them *Sinaistic* lecturers; they emit unambiguous truths, certainties as unchallengeable as the numbered Commandments. (Their style comes perhaps closest to that which fits the classical theory of the lecturer, discussed earlier.) At their best they can hope to approximate Boltzmann, whose precise preparation and carefully structured delivery, tempered by his great humor and humanity, made his lecture room always overflow with students and visitors. More usually, however, such lecturers function negatively; they eliminate quickly any ambiguity of what there is to be looked at in the view field. Unresolved problems are rarely hinted at. Physics is physics, and that is that. Depending on their temperaments, the students may well be frightened or bored.

On the other hand there are lecturers who use, as it were, a projection lens with large depth of focus and of field, who develop to low rather than high contrast, who accentuate, consciously or unconsciously, the fringe areas that surround the more or less predominant arrow in the view field. In lecture, they may start out on a clean line of argument; but then they wander off in this direction or that, interrupt themselves perhaps, stand for a time in silent thought, and often leave the impression that the really interesting part of the explorations is just beyond the clearer part of the view field altogether. One might call a lecturer of this sort excursive or *exegetic*. In the extreme case, the student begins to wonder whether the main arrow in the view field is worth looking at at all.

Figure 3. The Lecturer: Sinaistic Reducer. (Reproduced by G.C. Sellery, The Renaissance, *University of Wisconsin Press, 1950, p. 197.)*

It should not be inferred that these two styles and the two mentioned before exhaust the list of individual types. Moreover, mixtures are possible. Figure 3 shows a Renaissance woodcut of a classroom, clearly representing in the person of the lecturer a Sinaistic Reducer—a mixed type, protected by his desk and "Silence" sign and hewing closely to the book, which his students clearly are meant to accept without debate as the Truth.

Unplanned Communication

This attempt at a theory of what lecturers do has led us to consider the importance of the secondary or fringe areas in the field that they project. In the fringe field, the subvocal, gestured, subliminal communication can be strong, perhaps unintentionally even a predominant part of the message. The efficacy of unintended communication is sometimes utterly astonishing.

One does not have to remember Sherlock Holmes' and every fortune-teller's gift for perceiving remarkable detail from minute evidences, or the reputed ability of some dowsers somehow to find evidences for underground water from the meager cues offered by the surroundings. Subvocal, gestured, and subliminal communication is, after all, a matter of everyday experience. Our capability to discern a great deal from the spontaneous part of behavior is a generally operative factor.

A useful distinction between "formal" and "informal" learning has been made by anthropologists. Informal communication also occurs between the lecturer and the class. It is, of course, difficult to be precise and quantitative about these matters, and therefore they are undoubtedly not congenial to most scientists. Nevertheless, we should be prepared to find in this sort of study important clues for characterizing those lecturers who make a great deal of difference to the student. Indeed, again from a purely evolutionary argument, our sensitivity to the presence or absence of important small clues in the subliminal range is very plausible. Survival often depends on the recognition of small signals before they have grown to their full amplitude. So, too, children must learn to respond to voice elements and gestures long before they understand the meaning of language and social customs and thereby communicate with their guardians. I have little doubt that these subliminal elements also play a large role in determining the difference between a good lecturer and one who is poor as a teacher though competent as a scientist.

One may well expect that research on this subject will yield interesting results and perhaps, by making lecturers more aware of what kind of signals they are really transmitting, improve the educational process.

Demonstrance as Ritual

These considerations bring us finally to the last of the functions which I have tried to distinguish in the performance of the demonstration by the live lecturer. I have emphasized that a demonstration is a creative act having some pedagogic functions that cannot otherwise be fulfilled as well or at all (e.g., in offering contact with high-order reality; in showing something beautiful, in its proper setting, to the proper time scale). Second, the lecturer, the audience, and the material form a coupled system, one whose coupling from moment to moment can be adjusted by the live lecturer to be properly loose rather than hypnotically tight. Third, the lecturer, in personal participation, emphasizes the importance of the subject by a willingness to incur risks of various sorts and degrees. Fourth, personal participation on the part of the lecturers or demonstrators conveys to the student subliminal but important cues concerning the subject and its practitioners, cues that appear both in the free motion of the mind and in the manner in which the equipment is handled. Fifth, the attention lecturers give to the demonstration equipment directs the attention of their audience away from person-to-person relationships and serves the audience as a model of the person-subject or person-thing relationship that is basic to the scientific approach.

But one must say at least a brief word about the fact that the lecture is also a ritualistic activity, like any meeting that has a gathered audience on one side and a person performing before them on the other. Of course, it can be, and all too often is, *merely* a ritual, even if it is interspersed with demonstrations. As the anthropologist Bronislaw Malinowski asked about certain utterances, we may wonder, are some of these lectures "used primarily to convey meaning? Certainly not! They fulfill a social function and that is their principal aim, but they are neither the result of intellectual reflection nor do they necessarily arouse reflection in the listener." Such lectures have mainly the moralizing purpose that attached to the ceremonial public dissections still performed in England in the seventeenth century, remnants from the Middle Ages.

But "ritual" should not in itself be dismissed, even by physicists. On the contrary, ritual is a universally operating mechanism for social expression and function. For example, the externalization of attention from the lecturer's person to some object at a moment when both the demonstrator and the audience together take it very seriously is a general, deeply important phenomenon. It is invoked again and again for communicating at one and the same time knowledge, affect, and a sense of social cohesion of the group—from the waving of a flag to the display of the monstrance during Mass. The experimental demonstration apparatus, held before the audience or made to operate before it, has in addition to its didactic purposes the symbolic function of stressing the importance of the occasion. (It is with this in mind that I used earlier the word *relic* in describing a certain type of demonstration apparatus that has historic meaning.) And this is by no means the only ritualistic function. One other that must be evident to everyone is the catharsis that the audience experiences when a well-managed demonstration suddenly resolves a posed problem.

In looking back over these reflections on the functions and styles of demonstration, I am again struck by the number and the variety of investigations that should be taken up by those concerned with educational research. Needless to say, neither this essay nor a report on experimental results can hope to tell any individual lecturer directly and immediately what practical changes in style and manner he or she should make. There is no common mold for all. One can do only what is appropriate to one's temperament and training. And even that one learns only very slowly.

But one can become better aware what one's own style and traits *are*, and so take advantage of one's own best potential. And in the long run, one can expect to learn of possible improvements in the art of teaching over its present state. We may perhaps also find in such a study the answer to a curious puzzle: why even the hastily prepared and only half-visible demonstrations in so many of our science courses are, after all, sometimes remarkably effective.

References

1. *Advice to a Lecturer* (London: Royal Institution [1960 reprint]). The brief pamphlet is an anthology largely taken from *The Life and Letters of Faraday*, by Bence Jones (1870). Faraday entered the Royal Institution as a laboratory assistant in March 1813, and for over thirty years, "was the foremost scientific lecturer in London or indeed the whole of England" (Ibid., p. 1).
    From the large more recent literature on lecture demonstrations, I would select, first of all, the following two books: Harry F. Meiners, ed., *Physics Demonstration Experiments* (New York: Ronald Press Co., 1970), Vol. 1, Chs. 1–4; C. Taylor, *The Art and Science of Lecture Demonstration* (Bristol, England: A. Hilger, 1988).

2. E. Smith, "Early History of the High School," *Annual Report of the School Committee, Cambridge, Massachusetts (1892)*, p. 54. Cited by S. Rosen in *Am. J. Phys.* 22:194 (1954).

3. On beginning his career as mathematics teacher in Graz, Kepler, in part because of his poor delivery, "had only a few listeners the first year, and none at all the second" (M. Caspar, *Kepler*, New York: Dover, 1993), p. 56.
    Max Planck recalls about his student days: "I must confess that the lectures of these men netted me no perceptible gain. It was obvious that Helmholtz never prepared his lectures properly. He spoke haltingly, and would interrupt his discourse to look for the necessary data in his small notebook; moreover, he repeatedly made mistakes in his calculations at the blackboard, and we had the unmistakable impression that the class bored him as much as it did us. Eventually, his classes became more and more deserted, and finally they were attended by only three students; I was one of the three..." (Max Planck, *Scientific Autobiography and Other Papers*, trans. Frank Gaynor, Philosophical Library, New York, 1949, p. 15).

4. P. A. Schilpp, ed., *Albert Einstein: Philosopher-Scientist* (Library of Living Philosophers, Evanston, IL, 1949), p. 15.

5. Sheila Tobias, "Peer Perspectives on the Teaching of Science," *Change*, vol. 18, no. 2 (March/April 1986), pp. 36–41.

6. For example, see C. Ramsauer, *Grundversuche der Physik in historischer Darstellung* (Springer, Berlin, 1953), Vol. I, p. 169: "In 1886, it occurred to Hertz, while using a pair of Riess coils for lecture demonstration purposes, that a very weak discharge in one of the coils is enough to cause sparks in the other..." This was the beginning of his work on electromagnetic waves.

7. There are other good reasons for preserving historically significant scientific equipment; they are given in the pamphlet on this topic, obtainable gratis by writing to the Director, Center for History of Physics, American Institute of Physics, 1 Physics Ellipse, College Park, MD 20740.

8. Jacques Hadamard, *The Psychology of Invention in the Mathematical Field*, (Princeton, NJ: Princeton University Press, 1945), Appendix II.

Acknowledgments

I wish to thank the original publishers for permission to convert the essays to their new form. The following list gives the primary sources on which the essays in this book were based; in each case, the material was reworked for this publication.

EINSTEIN'S INFLUENCE ON THE CULTURE OF OUR TIME is based on the introductory chapter, "Einstein and the Shaping of Our Imagination," in Gerald Holton and Yehuda Elkana, *Albert Einstein, Historical and Cultural Perspectives* (Princeton University Press, 1982).

EINSTEIN AND THE GOAL OF SCIENCE is based on the chapter, "Toward a Theory of Scientific Progress," in *Progress and its Discontents*, eds., G. A. Almond, M. Chodorow, and R. H. Pearce (University of California Press, 1982).

OF PHYSICS, LOVE, AND OTHER PASSIONS: THE LETTERS OF ALBERT AND MILEVA is based on an essay in *Physics Today*, vol. 47 (Part 1 August, and Part 2 September 1994).

WHAT, PRECISELY, IS "THINKING"?...EINSTEIN'S ANSWER is a new version of an essay which appeared in *The Physics Teacher*, vol. 17, no. 3 (March 1979).

WHAT PLACE FOR SCIENCE IN OUR CULTURE AT THE "END OF THE MODERN ERA"? is based on a paper delivered at the 1993 Sigma Xi Forum, *Ethics, Values, and the Promise of Science*, February 26, 1993 (published in the *Forum Proceedings* by Sigma Xi in 1993), and on a lecture given at the University of Vienna on April 20, 1993 (published in *Public Understanding of Science*, vol. 2, 1993).

THE PUBLIC IMAGE OF SCIENCE is adapted from chapter 14, "Modern Science and the Intellectual Tradition" in *Thematic Origins of Scientific Thought: Kepler to Einstein* (Harvard University Press, 1973).

"DOING ONE'S DAMNEDEST": THE EVOLUTION OF TRUST IN SCIENTIFIC FINDINGS is based on the Arthur Miller Lecture in Science and Ethics, given at the Massachusetts Institute of Technology on April 7, 1990.

IMAGINATION IN SCIENCE is based on a paper delivered at the Spoletoscienza Festival, Spoleto, Italy, 29 June 1991, and on chapter 7, "L'immaginazione nella scienza," in *La Responsibilitá della Scienza* (Rome: Gius. Laterza & Figli, 1993).

UNDERSTANDING THE HISTORY OF SCIENCE is based on a lecture in a series given during a lecture tour in China, 1985.

THE TRIALS OF J. ROBERT OPPENHEIMER is based on a review in *Partisan Review*, vol. XLVII, No. 3 (July 1981).

PERCY W. BRIDGMAN, PHYSICIST AND PHILOSOPHER is adapted from "A Personal View of P. W. Bridgman, Physicist and Philosopher," *Methodology and Science*, vol. 26–01 (1993).

I. I. RABI, CITIZEN-SCIENTIST is based on a review in the *The Boston Sunday Globe* (17 May 1987).

FEYNMAN'S ADVENTURES is an expanded version of an essay which appeared in the *The Boston Sunday Globe* (17 March 1985).

MICHAEL POLANYI AND THE HISTORY OF SCIENCE has its source in a paper delivered at the Polanyi Centennial Conference, Harvard Divinity School, February 28, 1992, and in an essay in *Tradition and Discovery*, vol. 19, no. 1 (1993).

HOW CAN SCIENCE COURSES USE THE HISTORY OF SCIENCE? is adapted from an invited lecture given at the annual meeting of the American Association for the Advancement of Science, 14 February 1993, in the session on "History and Philosophy of Science as Aids to Science Teaching."

FARADAY'S "ADVICE TO A LECTURER," UPDATED is based on "What Is Conveyed by Demonstrations?" in *Physics Demonstration Experiments*, vol. 1, ed., Harry F. Meiners (New York: Ronald Press, 1970).

Index

About the Author

Gerald Holton, as scientist, historian of science, and educator, has devoted himself to the development of a framework for understanding the cultural role of modern science and, where possible, to making such understanding count in policy decision.

Having obtained his doctorate in physics in P. W. Bridgman's high-pressure physics laboratory at Harvard, he has remained at the university as Mallinckrodt Professor of Physics and Professor of the History of Science. While continuing research on the molecular structure of materials under pressure, he signaled his developing interest in the history of science first in his widely used textbook, *Introduction to Concepts and Theories in Physical Science*. Originally published in 1952, and remaining in print in a revised edition (Princeton University Press), it became also the basis of the national curriculum development project resulting in *The Project Physics Course*.

Among his scholarly publications in the history and philosophy of science, the ones perhaps best known are his pioneering researches on the work of Albert Einstein and other twentieth-century physicists. These are to be found in his books, including *Thematic Origins of Scientific Thought: Kepler to Einstein* (Harvard University Press), *The Scientific Imagination: Case Studies* (Cambridge University Press), *The Advancement of Science, and Its Burdens: The Jefferson Lecture and Other Essays* (Cambridge University Press), *Science and Anti-Science* (Harvard University Press), and *Albert Einstein: Historical and Cultural Perspectives* (Princeton University Press, with Y. Elkana).

Professor Holton received the George Sarton Medal from the History of Science Society, and served as president of the Society. Among his other honors are the Andrew Gemant Award of the American Institute of Physics, the Millikan and Oersted Medals of the American Association of

Physics Teachers, and the selection by the National Endowment for the Humanities as its tenth Jefferson Lecturer. He has given the Albert Einstein Lecture at Rockefeller University, the Herbert Spencer Lecture at Oxford University, and was exchange professor at Leningrad University.